Edward Enfield

Indian Corn

Its value, culture, and uses

Edward Enfield

Indian Corn
Its value, culture, and uses

ISBN/EAN: 9783744742948

Printed in Europe, USA, Canada, Australia, Japan

Cover: Foto ©berggeist007 / pixelio.de

More available books at **www.hansebooks.com**

INDIAN CORN;

ITS

VALUE, CULTURE, AND USES.

BY

EDWARD ENFIELD.

NEW YORK:
D. APPLETON AND COMPANY,
443 & 445 BROADWAY.
1866.

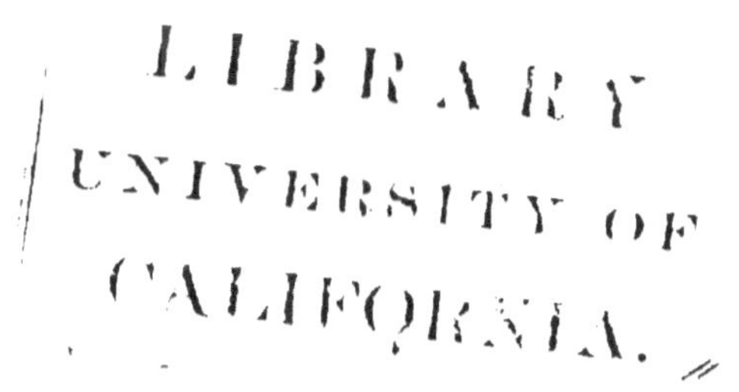

TO THE EDITORS AND LEADING WRITERS OF THE AGRICULTURAL PRESS.

As a feeble tribute of admiration, Gentlemen, for your valuable services in advancing the great farming interest of our country, the author begs leave to inscribe to you this humble effort.

Agriculture is the acknowledged basis of our national growth and prosperity. It has contributed, more than any other cause, to make our country what it is, and is destined to be equally instrumental hereafter in making it all that it promises to be.

But while we all perceive and readily acknowledge the great national importance of this branch of industry, should we not equally recognize the vast and beneficent influence exerted by the class of writers I am addressing?—a class, numerically small but influentially potent, who, by advancing our agriculture, have contributed more to develop our material wealth and power than any other equal number of men in the country. No man who has paid any attention to the progress of American husbandry during the last few years, and to the direct influence exerted upon it by the class of periodicals especially devoted to it, can fail to realize how much the country is indebted to the conductors and writers of such journals.

Wherever these sheets have penetrated the rural districts, the effect has been immediately obvious, in the ameliorated condition of the soil, in the improved quality and augmented quantity of farming products, and in the general thriftiness, the social and moral advancement of the farming population.

It is not the mere language of compliment, Gentlemen, to say that, while you have been steadily, and patiently, and zealously engaged from month to month, and from year to year, in writing up our farmers to a higher level of intelligence and success, you have at the same time, and in the same measure, been writing up to a higher level the prosperity and affluence of our common country.

The clever author of "Ten Acres Enough," in accounting for the success of his farming enterprise, remarked, with pardonable complacency, that he had *manured his soil with brains.* The metaphor will bear a wider application. It may be said with equal propriety that our agricultural writers have been for a series of years manuring a continent with the same remarkable fertilizer.

It is one of the most auspicious signs of the times, that the general public are beginning to take a much livelier interest than ever before in all that relates to the cultivation of the soil. Horticultural magazines and farming journals are finding their way into hundreds of families who, having no ground to cultivate, are yet waking up to a general interest in the subject. Quotations from the agricultural press are now frequently and almost constantly seen in the general newspaper; and people are beginning to discover that husbandry, in one form or another, is related to every condition of life, and that the welfare of the whole community is bound up in the success and prosperity of the farmer.

To you, Gentlemen, we are largely indebted for this improved and encouraging condition of the public mind. And though your services in this great cause have never yet been adequately appreciated, the day is undoubtedly near when a more generous recognition will be accorded to the influence and usefulness of your labors. One thing, at least, is certain. If contemporary justice is not rendered to the leaders and guides and expounders of American agriculture, another generation will gratefully record their names among the benefactors of our country.

I am, Gentlemen, respectfully and gratefully yours,

THE AUTHOR.

ACKNOWLEDGMENTS.

In the preparation of this work, the Author has derived valuable information from various sources, which it gives him pleasure to acknowledge. Where the language of another writer has been employed, it is duly credited in the context. Besides these instances, he is indebted for facts and opinions to the following authorities:

" Johnston's Agricultural Chemistry," " The American Farmer's Encyclopædia," "Burr's Field and Garden Vegetables of America," "Harris's Rural Annual," and " Tucker's Illustrated Annual Register." Also to the productions of Dr. Harris and Dr. Fitch, " On Injurious Insects; " to the " Transactions of the New York State Agricultural Society," and to the American Institute Farmer's Club, whose weekly discussions abound in valuable practical information.

Prominent also among the works that have been of service to the writer, are the Agricultural Journals of our country. While they are gratefully recorded here as valuable auxiliaries in the present undertaking, the record may, perhaps, prove serviceable to the farming community by attracting their attention to these fruitful sources of knowledge and sure guides to prosperity.

The American Agriculturist...........New York City.
  " *Weekly Tribune*................   "   "

ACKNOWLEDGEMENTS.

The Country Gentleman............Albany, N. Y.
  "   *Working Farmer*................New York City.
  "   *New England Farmer*..........Boston, Mass.
  "   *Boston Cultivator*.............. "   "
  "   *Farmer and Gardener*..........Philadelphia, Pa.
  "   *Rural New Yorker*.............Rochester, N. Y.
  "   *Ohio Farmer*...................Cleveland, Ohio.
  "   *Massachusetts Plowman*.........Boston, Mass.
  "   *Prairie Farmer*................Chicago, Ill.
  "   *Farmers' Advocate*............. "   "
  "   *Wisconsin Farmer*.............Madison, Wis.
  "   *Maine Farmer*.................Augusta, Me.
  "   *Genesee Farmer**...............Rochester, N. Y.
  "   *Germantown Telegraph.* ....Germantown, Pa.
Colman's Rural World.............St. Louis, Mo.
The Western Rural.................Detroit, Mich.
  "   *Culturist*......................Philadelphia, Pa.
  "   *Rural American*...............Utica, N. Y.
  "   *Rural Register*................Baltimore, Md.
  "   *Iowa Homestead*................Des Moines, Iowa.
  "   *Southern Cultivator*...........Athens, Ga.
  "   *California Farmer*............San Francisco, Cal.

* Recently merged in the *American Agriculturist.*

PREFACE.

THE importance of the subject, and the absence of any work specially devoted to it, is deemed a sufficient apology for the appearance of this book. For a number of years the author has given much attention, both theoretically and practically, to the culture and uses of Indian corn, and has, during that time, accumulated a considerable amount of materials relating to the subject, and mainly derived from the experience of farmers in various sections of the country.

Since no abler pen has undertaken to supply a want widely felt and acknowledged in the agricultural world, he has at length concluded to digest and arrange his store of materials on hand into the form of the present volume, which is now offered to the public with a lively sense of its imperfections, but not without a profound conviction of the importance of the subject.

The aim has been to condense within a small compass all needed and useful information, and to state facts, opinions, and results, as clearly and concisely as possible.

In the discussion of some of the leading topics, the author would gladly have devoted more space, in proportion to their importance, but it was found that such a course would render the work more voluminous and expensive, thereby possibly excluding it from the largest circle of readers.

The critical reader is here notified that he will find, in the course of these pages, some repetition of the leading thoughts which it is the object of this book to develop and impress. When a topic, already once treated, has reappeared in a different connection, especially if involving a principle of some consequence, the writer has not hesitated to improve the opportunity of reaffirming such principle, and again urging it on the attention of the cultivator. The same ideas have thus been, in several instances, partially reproduced. If they shall appear to the agricultural reader as important as they have seemed to the writer, no further apology will be needed. The reader who looks for imperfections will easily find them; but faults which, like this, have their origin in the force of the writer's convictions, however they may displease the critic, will not, it is thought, offend the practical farmer.

CONTENTS.

INDIAN CORN.

INTRODUCTION.

It appears, from the census returns of 1860, that there were at that time, 3,381,583 farmers in the United States, which, by the ordinary ratio of increase, would make the present number not far from four millions; most of whom are, doubtless, in the habit of raising an annual crop of Indian corn. This, at least, is to be presumed, for the crop is so universally cultivated, and so essential to the husbandman, that those omitting it must be extremely few in number.

Allowing for these exceptions, and for the interruptions resulting from the war, it may be taken for granted that there are, in round numbers, not less than three and a half millions of proprietary farmers engaged in the cultivation of this grain; some on fields measuring hundreds of acres, and some on limited patches of a few square rods; some producing fifteen or twenty bushels to the acre, and others one hundred and fifty or more; but all contributing to the grand result, and swelling the aggregate crop of the nation to such vast proportions as the world has never before witnessed.

Here, then, are two distinct objects brought to the notice of the reader, viz. :

The great staple crop of the country, and

The class of men engaged in producing it.

To the former of these topics the present volume is devoted. To the latter, let us accord the passing tribute of a few lines.

There are various reflections that give weight and consideration to the large and respectable body of men devoted to agricultural pursuits. The very nature of their occupation renders it of vital importance to the welfare of the community. The products of agriculture embrace articles of such indispensable necessity, that the continued existence of our population is literally suspended upon the tillage of the earth. The farmer feeds the community, and every member of it is thus daily, and almost hourly, reminded of his value and importance in the social scale.

But without dwelling on general considerations, it is sufficient to refer to a few prominent facts. It will be seen, from the census returns above referred to, that in 1860, the whole number of persons in the United States engaged in manufactures and kindred branches was 2,017,653; and of those engaged in commerce and connected pursuits, 757,773; while the number engaged in agricultural operations, as stated above, was 3,381,583.

Thus it appears that the farmers not only outnumber the merchants and the manufacturers, taken separately, but they surpass the combined numbers of those classes by more than half a million. It also appears that, by the ordinary ratio of increase, the number of farmers in the whole country, at the period of the next census, will probably exceed five millions, counting the heads of families merely, and not their dependents.

It is quite apparent, therefore, that this class of our citizens, unconspicuous as they have been in the retirement of their rural homes, have yet grown to dimensions, and risen to an importance, well calculated to arrest attention. But while their numbers are rapidly advancing, their achievements do not flag. The annual fruits of their industry, increasing with their population, have reached a prominence and magnitude everywhere seen and felt, and everywhere acknowledged to be without a parallel. American husbandry has made its mark in the world, not only by the intrinsic value, but equally by the quantities of its products. The unexampled amounts of grain and provision which it has annually poured into the channels of commerce, have justly challenged the attention and the amazement of mankind.

In whatever light we view this subject, we cannot fail to be impressed with the valuable services and the growing influence of our yeomanry. It is not a mere metaphor, nor even an exaggeration, to say that the destiny of the nation is in their hands. The national census is the history of their achievements and the monument of their greatness. Their position and influence in the community is a simple matter of fact which it is proper to recognize, legitimate to account for, and may be useful to contemplate, and which there can be no reason to ignore.

If the mere statement of these facts affords them any ground of complacency and self-gratulation, so does it also bring with it momentous responsibilities. To remind them of these is no idle compliment, but may serve a useful purpose. If they have done so much for their country in the past, what may they not do in the future?

The present is an eventful and auspicious epoch in our history, holding out to our people, and especially to our farming population, great and glorious opportunities. We stand between a dreary past and a hopeful future. Having extinguished, with a rapidity and completeness unexampled, the most stupendous rebellion on record; having continued through the whole of that struggle to exhibit and unfold with scarcely any interruption our immense material resources; having made that fiery tribulation the occasion and opportunity for developing an amazing national vitality, a physical energy, a force of character, and a moral power surpassing our own previous conceptions, and scarcely yet credited by the rest of the world; having confirmed and established in the reluctant confidence of foreign nations, the vigor, efficiency, and permanency of our government; having thrown open our vast domain of fertile acres to the people of all climes, thus offering a bid for population beyond the competition of other powers; having invited, facilitated, and secured a steadily increasing tide of immigration from abroad, it would certainly appear as if the era upon which we are now entering holds out a prospect beyond any thing hitherto revealed to mankind. We stand on the threshold of a future so full of promise, so radiant with hope, so teeming with possibilities and opportunities, that imagination can scarcely overdraw, nor enthusiasm exaggerate the approaching scenes of prosperity, affluence, and power.

To you, Brother Farmers, such reflections as these cannot be without interest, for with you it mainly rests to realize for your country these well-founded and rational anticipations. You hold the keys that shall unlock the treasures of the earth. In your hands are the magic wands that shall convert prophecy into history, and organize possibilities into accomplished events, transmuting the visions of the future into solid facts, and crystallizing anticipated scenes into living realities.

To you, then, gentlemen, may the writer be allowed to address a few plain and candid remarks.

If the prosperity of this nation is founded upon the prosperity and success of its farmers, then arises at once the vital question, On what does the success of the farmer depend? The obvious answer is, that it depends mainly upon his getting from his land the largest amount of products, at the lowest rate of expense. To do this requires not only industry but intelligence; not merely the faculty of working, but the faculty of thinking. The man who, by combining thought with action, contrives to get, year after year, five or six bushels more of wheat, and ten or fifteen more of corn from an acre of ground than his neighbor gets, under like circumstances, will undoubtedly, other things being equal, outstrip his neighbor in the race of prosperity. If this is true in reference to individuals, it is equally so and the effect is far more striking in reference to communities.

Let us take, for example, the corn crop of the United States, and see what the difference would amount to, in the aggregate, if every farmer in the country, at the period of the last census, had raised, with little or no additional expense, five bushels more to the acre. This result was not merely possible, but easy to accomplish, and would have made a net addition of nearly one hundred and thirty million bushels of corn to the product for that year. This being the difference on one crop out of a dozen or more, we may form some idea of the total excess that would result, in a single season, from even a small increase all around in the ratio of production.

Now here, gentlemen, is the point which ought to arrest your attention. The average yield per acre, throughout the country, is entirely below what it should be. The product of Indian corn might just as well be, on a general average, fifty bushels to the acre as thirty or thirty-five; and in putting the amount at fifty bushels, the standard is still too low.

It is easy, however, to perceive, and is well understood, that the rate of yield here complained of is the fault of a part of the agricultural community only, and not of the whole; and it is but just to remark, that low as this average appears, it is nevertheless above that of former years, and has been slowly, and with some fluctuation, gaining ground for nearly half a century. It must also be admitted, and is entitled to be considered, that notwithstanding this low rate of production, the aggregate amounts of our various crops have risen to proportions truly amazing, and have, as already stated, contributed immensely to the growth and power of the country.

But after all these admissions, though in looking at the grand aggregates, we find them, in comparison with former years, steadily advancing, and though we find the broad result to be national development and prosperity beyond that of any other people, still the inquiry arises, and forces itself upon the mind, What would have been, or rather, what might not have been accomplished, with a larger average yield? What other, and higher, and more incredible results might not have been achieved, had the ratio of production been fifty bushels per acre for corn, with a corresponding increase for all other crops?

Now, to every cultivator of the soil this question of acreable product is one of no little moment; and he has already gone far toward solving it, when he has committed his grain to the ground in the spring. It is indeed a serious question, not only to himself but to the community as well, whether he shall gather twenty bushels from an acre or one hundred and fifty, or what intermediate number he shall reach between these extremes. One thing at least is certain: in the present state of intelligence, with the existing facilities and recently improved methods of culture, no man of ordinary enterprise will be satisfied with any such quantity as the average yield of the last decade. It cannot be denied that thirty-three bushels per acre is too low an average for the whole country, considering that one hundred bushels are by no means unusual, and that much higher figures have been reached, even all the way up to two hundred bushels.

Whatever has been done in repeated instances, by various parties and under differing circumstances, is surely a reasonable standard for every man to aim at, and one which no true farmer will permit himself to lose sight of. Knowing the limit of possibility, it is only necessary to know further what are the conditions essential to its attainment. Comply with these, and you achieve the result. Let every farmer make up his mind, at planting, how many bushels per acre are fairly within his reach. Let him fix his mark in the spring, with a firm resolve to come up to it. He who determines to achieve whatever has been proved reasonably possible, may safely aim at an elevated mark; and if he conforms to the laws of reason, and nature, and common sense, will hit the centre of his target at every shot.

But there are, gentlemen, two great agencies operating throughout the country, the tendency of which is so favorable and so powerful for good, that I cannot forbear to urge them on your attention. I allude to the influence of farmers' clubs and farming journals. No man engaged in agricultural pursuits can expect to keep up with the spirit of the times, without availing himself of these useful and invaluable means of improvement. If every man who wins his livelihood from the soil, would appropriate the experience of his fellow-cultivators by connecting himself at once with a farmers' club, and subscribing promptly to an agricultural journal, causing it to be taken and read in his family, the effect on the soil and crops of the ensuing season would be marvellous and magical all over the country.

The valuable facts and experiments, and the variety of information which abound in these journals, produce their legitimate results, in improving, elevating, and enriching the farmer, with just as much certainty as does the manure applied to his crops, or the tillage bestowed on the soil. The conductors and writers of this branch of the press devote themselves with untiring industry to collect and disseminate the opinions and experience of our wisest practical men, and the scientific principles laid down by the highest authorities.

It is not easy to determine how many of these journals are at present taken and read throughout the country, but it seems probable that the number of subscribers, putting all the journals together, would not much exceed one-third of a million, which is less than one man in ten of the agricultural proprietors, and scarcely one in forty of the farming population. It must be admitted that this ratio of readers to the whole number of cultivators is discreditably low. In an agricultural community numbering four million families, there ought to be, at the least calculation, one million subscribers to this class of periodicals; nor is it easy to assign any reason why this number should not yet be reached before the period of the next general census. We should then have three reading farmers where we now have one, and the effect upon agriculture which such an increase of intelligence would everywhere produce it is scarcely possible to overrate.

It rests with you, brother farmers, to introduce this new era of diffused intelligence, by doubling or tripling, as you easily may, the circulation of the agricultural press. Should you enter thoroughly into the spirit of this subject, the purpose would be accomplished. You would thereby change the aspect and condition of fields and farms all over the land, imparting to every meadow a brighter green, and to the fruits of autumn a deeper tinge of gold. You would communicate ideas to ploughshares, convert the hoe into a calculator, and endow the spade with thought.

What effect this would produce upon the future grain crops of the country, it is not difficult to perceive. Even without counting any increase from this cause, the corn crop for 1870, as will be seen by the estimate on another page, is likely to exceed a thousand million dollars in value. The grain itself, according to that estimate, will be sufficient to feed not only our own people, but half the population of Europe in addition, for more than twelve months; while the money value of such annual crops would, in the course of three years, suffice to extinguish our national debt, and leave a balance in the treasury.

It seems to me, Farmers of America, that such a

record will be the best possible commentary on the Great American Rebellion, and the best possible rebuke to the numerous tribe of croakers and prophets of evil abroad, who have so long and steadily been gloating over the approaching dissolution of our Union.

That the citizen soldiers of this country, after bringing to a successful close a civil war so formidable and terrific, should have laid aside promptly, in the very hour of triumph, the arms which they had covered with glory, and gone back quietly to their cherished homes, and to the beneficent occupations of peace; that a class of men notoriously ardent and susceptible should abandon at once and with complacency, the exciting scenes of martial life, and the fields of all their fresh renown, satisfied with a sense of duty performed and a country saved; that so soon after turning their backs upon the field of battle, they should exhibit to the world a countless array of harvest fields stretching over a thousand hills and valleys, and covering a land redeemed by their valor and now embellished by their toil—this indeed is a moral spectacle instructive to the world, and more to be prized than all the material prosperity and affluence which it indicates.

EXTENT AND VALUE OF THE CORN CROP.

General View.—The extent of the corn crop of this country, and its importance in an economical and commercial view, have risen to a scale of magnitude that overshadows all other crops. It appears, from the census of 1860, that the corn crop of that year was over eight hundred million bushels, while the product of wheat, rye, oats, barley, buckwheat, peas, beans, and potatoes, taken in their entire aggregate, was less than that of Indian corn by more than three hundred million bushels. Compared with the wheat crop alone, the product of corn is very nearly five times greater; and when the comparison is extended beyond our own country, it is found that the corn crop of the United States is about equal to the wheat crop of the whole earth.

The following are the decennial returns of Indian corn, as given in the census tables of the last three decades:

<pre>
For 1840....................377,431,874 bushels.
    1850....................592,071,104   "
    1860....................838,792,740   "
</pre>

It appears, from this comparison, that the increase from 1840 to 1850 was nearly two hundred and fifteen million bushels, and from 1850 to 1860 it was nearly two hundred and forty-seven million bushels. For the entire period of twenty years, the gain was over four hundred and sixty-one million bushels, being at the rate of a little over six per cent. a year, or sixty per cent. for each decade.

The following table exhibits the corn crop of 1860 in comparison with some of the other leading crops of the country:

Corn	838,792,740	bushels.
Wheat	173,104,924	"
Rye	20,976,285	"
Oats	172,554,688	"
Barley	15,635,119	"
Buckwheat	17,664,914	"
Peas and Beans	15,188,013	"
Potatoes	110,571,201	"

The aggregate number of bushels for these eight crops is thirteen hundred and sixty-four million, four-hundred and eighty-seven thousand, eight hundred and eighty-four, making an average of over one hundred and seventy million bushels for each crop.

The returns of the corn crop for the several States and Territories for 1850 and 1860, are indicated in the following table, in which the States are arranged in the order of the alphabet and not in the order of their yield.

	1850.	1860.
	Bushels.	Bushels.
Alabama	28,754,048	33,226,282
Arkansas	8,893,939	17,823,588
California	12,236	510,708
Connecticut	1,935,043	2,069,835
Delaware	3,145,542	3,892,337
Florida	1,996,809	2,834,391
Georgia	30,080,099	30,776,293
Illinois	57,646,984	115,174,777
Indiana	52,964,363	71,588,919
Iowa	8,656,799	42,410,686
Kansas		6,150,727
Kentucky	58,672,591	64,043,633
Louisiana	10,265,273	16,853,745
Maine	1,750,056	1,546,071
Maryland	10,749,858	13,444,922
Massachusetts	2,345,490	2,157,063
Michigan	5,641,420	12,444,676
Minnesota	16,725	2,941,952
Mississippi	22,446	29,057,682
Missouri	36,214,537	72,802,157
New Hampshire	1,573,670	1,414,628
New Jersey	8,759,704	9,723,336
New York	17,858,400	20,061,049
North Carolina	27,941,051	30,078,564
Ohio	59,078,695	73,543,190
Oregon	2,918	76,122
Pennsylvania	19,835,214	28,196,821
Rhode Island	530,201	461,497
South Carolina	16,271,454	15,068,606
Tennessee	52,276,223	52,089,926
Texas	6,028,876	16,500,702
Vermont	2,032,396	1,625,411
Virginia	35,254,319	38,319,999
Wisconsin	1,988,979	7,517,300
Territories	440,540	2,388,147
	592,071,104	838,792,740

The principal corn-growing States rank for 1860 in
the following order :

1. Illinois.	6. Tennessee.	11. N. Carolina.
2. Ohio.	7. Iowa.	12. Mississippi.
3. Missouri.	8. Virginia.	13. Pennsylvania.
4. Indiana.	9. Alabama.	14. New York.
5. Kentucky.	10. Georgia.	

The first six of these States produced in 1860 about four hundred and fifty million bushels, being more than half the product of the whole country.

In 1840, Tennessee was the greatest corn-producing State; in 1850, Ohio took the first rank, and in 1860 Illinois stood at the head.

The greatest gain made by any of the principal corn-growing States has been made by Iowa. In twenty years the product of that State has increased from less than one and a half million bushels to over forty-two million bushels.

The proportion of Indian corn to the whole number of inhabitants is not a little remarkable. Compared with that of potatoes and wheat, it stands as follows:

```
Potatoes to each inhabitant,.............. 210 lbs.
Wheat,      "          "      .............. 330  "
Corn        "          "      ............1,590  "
                                          -------
                                           2,130
```

This gives an aggregate of more than two thousand pounds of food to every man, woman, and child in the country, from three leading crops.

The following is an approximation to the average yield per acre, and the number of acres in corn, for the last two decades:

2

	ACRES IN CORN.	AVERAGE YIELD.
		Bushels per acre.
1860......................	25,417,961	33
1850......................	23,682,844	25
Increase..............	1,735,117	8

MONEY VALUE OF THE CORN CROP.—In estimating the value of this crop, it is to be remembered that the market price of corn varies greatly between the East and West. In the city of New York it has ranged, during the last six years, from sixty cents up to two dollars per bushel, averaging during the last three years about one dollar and ten cents. At the West it has ranged much below these figures, probably from fifty to seventy per cent. lower; but as most of the corn in that section is consumed on the land where it grows, paying the farmer much better, on an average, than the market price, it is not easy to determine what the crop actually realizes to the producer. Taking into consideration, however, the various forms in which it is turned into money, and the range of market prices, it may safely be assumed that the corn crop brings, on an average, not less than sixty cents per bushel.

But there is an important item which, though it has found no place in the tables of the census, cannot properly be omitted in computing the product of Indian corn. It will be found that the stalk crop of the country, including all the stover of corn raised for all purposes, amounts to about forty million tons.* There is no regular market price established for this stover,

* See Estimate on page 177.

but its positive pecuniary value is not, for that reason, any less. It is variously estimated from three or four dollars a ton up to twelve dollars and over. In some parts of the country, and by many of the best farmers, it is considered quite equal in value to good hay.

As there is, however, some difference of opinion in regard to the value of corn-stalks, we will assume that they are worth five dollars a ton, on an average; although it is demonstrable that, when turned to the best account, they can be made to realize, in most cases, nearly or quite double that amount.

Taking the grain, then, at sixty cents per bushel, and the stover at five dollars per ton, the total value of the corn crop for 1860 will foot up as follows:

```
838,792,740 bushels of grain, at 60c......$503,275,644
40,000,000 tons of stalks, at $5..........  200,000,000
                                           ─────────────
                                            $703,275,644
```

ESTIMATED CROP FOR 1870.—In forming any conclusions on this subject, there is perhaps no better guide than the comparative increase of the crop during the last two decades. Though agricultural operations have been temporarily interrupted in a portion of the country by the events of the war, it is now probable that the nation will be soon restored to a condition of more than former prosperity; that whatever the country has lost by the Rebellion in agricultural products, will be more than compensated by the increased activity of the coming years; and that the census of 1870 will show that our staple crops have not lost ground in consequence of the war.

The increase of the corn crop during the twenty years from 1840 to 1860, was at the rate of a little more than six per cent. a year. It may then, we think, be fairly taken for granted, that the gain for the present decade will be, at least, equal to five per cent. a year. According to this ratio of increase, and taking the same valuation as before, the corn crop for 1870 will show the following aggregate, in quantity and value:

```
1,258,189,110 bushels of grain, at 60c....$754,913,466
60,000,000 tons of stover, at $5..........  300,000,000
                                           ─────────────
                                            $1,054,913,466
```

CONSUMPTION OF THE CROP.—In view of the present and increasing amount of this stupendous crop, it becomes an interesting and important inquiry, where and how it is consumed.

The amount of corn exported is small compared with that of wheat, and when viewed in contrast with the product of the entire crop, appears quite insignificant. The total exports of corn and wheat for the last six years, and the average per year, are as follows:

```
Corn, 40,895,237 bushels, average per year......6,815,872 bush.
Wheat, 112,938,693  "        "        "      ....18,823,115  "
```

Thus it appears that the ratio of corn exported is less than one per cent. of the whole crop, while that of wheat is very nearly eleven per cent., without including the shipment of flour, which during the same period averages 1,667,342 barrels per year. If this amount is added to the grain sent abroad, it will make the ratio of wheat exported about fifteen per cent. of the entire crop.

But there is another view of the export of corn which presents it in a more favorable light. While less than one bushel in a hundred is sent directly abroad, a much larger proportion than this is indirectly exported, in various forms, more remunerative to the farmer, and more profitable for the country. Indian corn enters, in a larger or less degree, into nearly all the beef, pork, mutton, butter, cheese, and lard produced by the entire farming community. These products are not only in great demand for domestic consumption, but are, all of them, with the exception of mutton, largely exported.

The beef shipped to Europe from the port of New York, during the last three years, amounts, on an average, to forty thousand barrels and fifty-four thousand tierces per year. The pork shipped during the same time exceeds one hundred and forty-seven thousand barrels on a yearly average, and other meats exported amount to over one hundred million pounds a year; while the aggregate of butter, cheese, and lard sent abroad during the same period is over three hundred and seventy-five million pounds. These results, however, are less than they would have been, in consequence of an exceptional decline in the export of provisions during the last year.

But far the largest consumption of Indian corn is by our own people. The home market, which is more easily reached, is vast in extent, and constantly increasing in its demand. Not only as a direct article of human food is this grain largely consumed here at home, but also, and to an almost incredible extent, as provender for the immense number and variety of our domestic animals. The same commodities to which corn contributes for export, it also produces or aids in producing on a very much larger scale for domestic consumption.

As an illustration of this, the quantity of beef, veal, mutton, and pork absorbed in a single year by the city of New York alone, is indicated by the following statement of live stock received for 1865 :

```
Beeves.....................................273,274
Veals...................................... 77,991
Sheep and Lambs............................836,733
Swine......................................573,197
                                         ─────────
    Total..................................1,761,195
```

Nearly the whole of this amount of animal food was consumed during the year, by the population of New York city and its vicinity; from which some conception may be formed of the quantity of meat required, and the quantity of corn used in producing it, for a population of over thirty millions.

The total amount of butter and cheese made in 1860 was about five hundred and seventy million pounds, and doubtless at the present time exceeds six hundred million pounds a year, most of which is consumed by our own people. In producing these articles, Indian corn is extensively employed, both the grain and the stover being found profitable for the purpose.

In a general view then, of the consumption of corn,

we discover how great a proportion of the crop is used for conversion into other kinds of food, and how largely it is fed out for this purpose on the land where it grows; thereby tending to increase the prosperity of the farmer by improving the quality of his soil. And herein consists one great advantage of this cereal over wheat. Though both are largely consumed at home, in one form or another, and both to some extent exported, yet the result in the two cases is very different.

The corn which the farmer converts into other products may be sent abroad or sold in any market without reluctance, and with advantage, for it leaves an enriched soil behind it, and brings back wealth to the country. But when the wheat crop is sold, whether at home or abroad, *an integral part of the farm is sold with it.* However largely it may be exported to Europe, still the land where it grew is despoiled without compensation, and the fertility of the earth is bartered for foreign gold. Already the deterioration of the soil resulting from this husbandry is, in some localities, severely felt, and farmers are anxiously looking around for new sources of fertility—for some adequate means of restoring to their land its departed virtue. But the system of special crops—of partial and exclusive husbandry, is wrong in principle, and should be reformed. If the practice of some farmers is continued, the loss to the country will in a few years be serious. If, for the sake of present gain, they continue to trade away the essential quality of their land, along with the grain it produces, selling out the very sources of their prosperity, the cream and essence of their farms, at sixty pounds to the bushel, it is certain to bring impoverishment to themselves or their children.

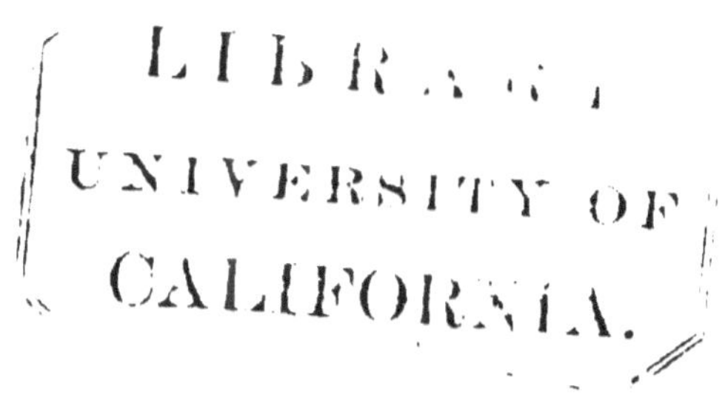

NAME AND ORIGIN.

Maize, or Indian Corn, is an herbaceous plant belonging to the family of grasses (*Gramineæ*). Its Botanical name, *Zea Mays*, is indicative of its nutritive quality, or power of sustaining life; the generic term, *Zea*, being derived from the Greek verb Zao, to live, while the word *Mays* is supposed to come from the Livonic Mayse, which signifies bread, or staff of life. It stands preëminently at the head of the cereals, or cereal grasses, which include all those that are cultivated for their grains, such as wheat, rye, maize, etc.; the term cereal being derived from Ceres, the name of the Pagan goddess that presided over grain and harvests.

In England, and on the Continent of Europe, the word *Corn* is applied equally to wheat, rye, and bread-stuffs in general; while in this country the use of the term is limited exclusively to maize. This specific application of the word has been confirmed by a judicial decision in Pennsylvania, in which it was ruled by the court that the word *Corn* is a sufficient description of Indian corn.

2*

ORIGIN.—In regard to the origin of this plant, although there has never been room for reasonable doubt, there have been those who fancied there was room for argument. America is clearly and beyond question the native country of Indian corn. Yet, from the commencement of its history, writers have not been wanting to contest this point, and to claim for it an Eastern origin. The weight of authority and of argument so entirely preponderates in favor of its American origin, that it is scarcely worth while, in a work aiming to be useful rather than learned, to waste the time of the reader with idle and unprofitable speculation

If any further evidence were wanting on this point, it may be found in the impossibility that a grain so nutritious, prolific, and valuable, so admirably adapted to the wants of man, could have existed in the Eastern world before the discovery of America without coming into general use, and making itself universally known. Had this cereal existed there at that period, it would have made its own record too clearly and positively to leave any doubt on the subject.

But on this, as on some other topics, there will always be found a class of minds ready to keep up an argument, whether there is any rational ground for it or not. It would seem to be time to dismiss the controversy by accepting, as final, the generally received conclusion, sanctioned by such names as Humboldt, Schoolcraft, and Prescott, that Indian corn was unknown to the Eastern world previous to the discovery of America.

But maize is not the only important plant indigenous to the Western world. Other vegetables highly prized, either for their usefulness or as luxuries, have had their origin here. Among these are included the *Tobacco plant*, and the *Potato*, both of which, but for the discovery of this continent, would still be unknown to the civilized world. Let all consumers, then, of these three important products, not forget their obligations to the immortal Genoese navigator, who, when he bequeathed a hemisphere to mankind, transmitted, at the same time, two priceless articles of food, and a weed of questionable value.

ADAPTATION TO VARIETIES OF SOIL AND CLIMATE.

The different conditions and qualities of soil resulting from the combination of its elements in varying proportions, are not only numerous, but probably incalculable. This diversity is strikingly illustrated in the fact that adjacent fields, however similar in appearance, are often found to differ, and sometimes widely, under the test of chemical analysis.

Yet of the almost endless diversity of soils, it is remarkable from how small a number maize is excluded. In nearly all of them it will grow to maturity, while in most of them it thrives with tolerable treatment, and repays a generous culture with an abundant crop. "Indian corn," says the *Farmer's Encyclopedia*, "can be cultivated on land, long after it has ceased to afford compensating crops of any other grain. It contends with poverty better than most other plants, and may be advantageously grown in any soil fit for cultivation, not excepting blowing sands or retentive clay."

"Corn will grow," says Mr. Joseph Harris, "on all soils, from the lightest sand to the heaviest clay,

among granite rocks and on the richest bottoms.   I
have been," he adds, "in a two hundred acre field
in Ohio that has produced annually a good crop of
corn for over fifty years without manure."

There is, indeed, scarcely a plant cultivated by man
that will grow with equal success in so great a diversity of soils.   The evidence of this fact is met with
in every direction through the country.   The travel-
ler whose way lies through cultivated districts, passes
over many qualities of land, yet nowhere does he miss
the ever-recurring cornfield.   However far he may
go, the soil along his way, like the landscape that
meets his eye, is constantly changing, but the crop of
growing maize continually reappears.   He passes a
thousand planted fields, so various in the composition
of their soils that scarcely any two of them are identical; yet of that thousand fields he finds a large
proportion planted with corn.

But though this ubiquitous cereal so readily adapts
itself to the new condition it finds in each new locality, making itself a home amid uncongenial elements,
and often growing with luxuriance where other cereals will scarcely grow at all, we are by no means to
infer that the quality of the land where it grows is a
matter of indifference.   On the contrary, there is no
grain more sensitive on this point than maize; none
that pays so munificently for fertility of soil in the
affluence of its yield.

Another property of this grain, which no other
cereal possesses in an equal degree, is the VARIETY OF
CLIMATE to which it is adapted, and the facility with

which it may be translated from one latitude to another.

Though originally found in or near the tropics, it has gradually extended beyond those limits, and may now be seen growing over the greater part of this continent, from about the fiftieth degree of north latitude to a corresponding parallel south, and extending to limits not far short of these in the Eastern hemisphere; though in the latter the growth is less vigorous and the maturity less certain. When transferred from one climate to another, if the distance be not so extreme as to render the contrast too violent, it gradually parts with the features and habits peculiar to its recent locality, and readily acquires those that are appropriate to its adopted home. By this beneficent arrangement of Providence its value and usefulness to man are greatly enhanced, not only by rendering the culture more general, but by affording the means of multiplying its varieties, improving its quality, and increasing its yield. Indeed, the important destiny for which this grain seems designed by the Creator, is in nothing more apparent than in the extensive area which it covers, and the variety of climes in which it thrives.

Though cultivated quite extensively and with considerable success in Southern Europe, as well as in portions of Asia and Africa, yet America seems to be its peculiar home, and the region of its highest perfection. From Maine to Oregon, from British America almost to the extreme verge of Patagonia, this legacy of the red man to the white, in some of its forms or varieties, is annually cultivated. Where frost-bound Minnesota lends to its growth a short and reluctant summer, where the rigor of a Canadian climate concedes to it a few weeks of glowing sun, or where the fervid sky of Kansas, or the sultry air and longer season of either Carolina produce an earlier development and a larger growth ; in short, wherever on this continent civilized man can exist with tolerable comfort, there will you find Indian corn pushing its little cylinder of folded leaves through the soil, or unfurling to the wind its long and graceful foliage, or lifting its newly formed tassel to greet the rising sun.

Though its growth under tropical skies is more rank and luxuriant, producing not unfrequently stalks of prodigious size, the yield of grain is found to increase as it advances toward the pole, and the largest product per acre is said to be obtained near the northern limit of its range.

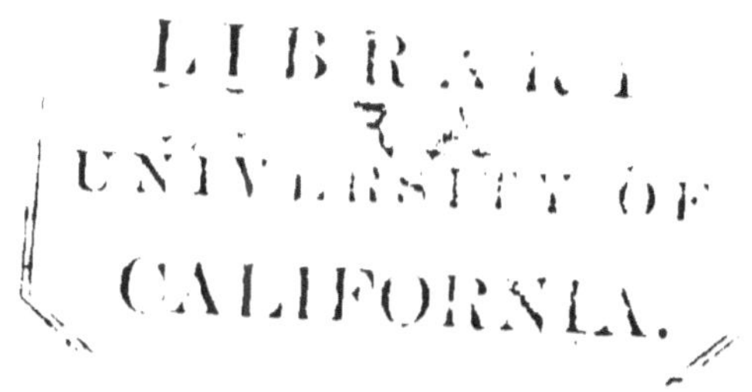

ADAPTATION TO THE WANTS OF MAN.

THE consumption of maize by the human family, and by nearly all domestic animals, has greatly increased within the last few years. As an article of food it is unsurpassed, and in the opinion of many unequalled, by any other grain or plant, combining, as it does, in suitable proportions, all the essential and valuable elements required for healthfulness and nutrition.

It appears from chemical analysis that Indian corn contains more oil and starch than wheat, with rather less gluten; and therefore, while scarcely inferior to that grain in nutritive value, far surpasses it, as well as the other cereals, in its fattening properties, which amount to nearly eighty per cent. of its composition. In point of nourishment it is second only to wheat, and even here the superiority of the latter is rather nominal than real; for if due allowance is made for the loss sustained by wheat in grinding and bolting, it will be found that a pound of corn yields quite as much nourishment as a pound of wheat. It is nearly fourfold more nutritious than the potato, which has so long been the great staple and staff of life with a numerous class, both in this and other countries ; and it has been proved by experiment that corn meal will sustain a workingman longer, when fed upon it exclusively, than any other grain.

The numerous preparations and manifold forms in which maize is fitted for the table, contribute to render it the most various and valuable, as it is with one exception the most abundant article of human food.

There is, however, a noticeable difference in the properties of the several varieties of this grain. While the constituents remain nearly the same in all, the proportions vary in which they are combined, and this fact still further increases its adaptation to the requirements of man and animals.

" For the colder half of the year," says the *American Agriculturist*, " the oil and starch of the corn are better adapted to the wants of the body, than the large amount of gluten in wheat. Corn contains all the elements needed in the body, and in just about the proportion they are required in winter, while they are nearly suited for food in warm weather."

The writer might have added with much truth, and making the case still stronger, that the Southern varieties, having a smaller proportion of oil than the flint corn of the North, are thereby rendered a softer and cooler food for the climate that produces them ; while the presence of a larger amount of vegetable oil in the maize of higher latitudes imparts to it the very quality that fits it for the region of its growth. It is found by travellers to the North that the larger the proportion of fatty elements contained in their food, the more easily they withstand the extreme severity of the temperature. Accordingly it appears that the seal, the bear, the water-fowl, and other animals that supply food to the natives of the frigid zone, acquire a superabundance of fat in the ratio of their proximity to the pole; and here we perceive the same law revealing itself in the vegetable kingdom. As man advances to the north, he finds the fuel that is demanded by the rigor of the climate partially supplied by the indigenous food that pertains to the latitude.

It is also to this peculiar property of maize that it largely owes its unrivalled excellence for fattening purposes. All domestic animals are easily and rapidly fattened when judiciously fed with corn meal; and, what is still more important, the flesh thus acquired is firmer and better than that produced by any other grain.

A further and more detailed consideration of the uses and value of this cereal for purposes of food may be found in a subsequent chapter.

CERTAINTY OF THE CROP.

INDIAN corn is usually accounted a certain crop, and in comparison with many others it undoubtedly is so. When seasonably planted, with due attention to the selection of seed, and tolerable care in the after culture, it has scarcely ever been known to result in failure. There are, of course, exceptional cases, arising from providential or human causes, such as unseasonable frost, absolute sterility of soil, utter neglect of the crop, etc. Apart from such instances as these, there is no seed which the husbandman commits to the earth with more certainty of securing some return for his labor.

Yet the difference between a moderate crop and a large yield is a very material point for the farmer to consider, though he too often overlooks it. Here is, in fact, the point where certainty ends and contingency begins. While he feels reasonably sure of a moderate yield, he is in danger of neglecting the means that would make him almost equally sure of a much greater one. The interval between thirty or forty bushels per acre and one hundred and fifty is very considerable, and if he allows himself to rest in the confidence of securing the former, he will be quite apt to lose sight of the possibility of the latter.

A small or moderate crop is nearly always a matter of tolerable certainty. But a large yield is encircled by elements of doubt. It is to some extent a question of sun and rain, of dew and frost, of tillage, fertilizers, etc. It is a question, too, about which squirrels and mice, and greedy birds, and myriads of voracious insects, have each a word to say.

Yet amid all these contingencies, and in the face of all these enemies, the intelligent husbandman reposes undismayed upon his conscious resources, reflecting that the same Providence that has strewed difficulties along his path has also endowed him with intellect and skill sufficient to counteract them. He goes into the cornfield with a clear head, a resolute purpose, and a strong faith, well provided with seed and implements, and with his favorite *agricultural journal*, and lo ! the formidable host of obstacles and enemies vanish from his presence ; and where a slovenly, unthrifty man, who never reads and never grows wiser, would possibly produce a crop of twenty or thirty bushels per acre, he, the intelligent farmer, raises one hundred bushels or more.

AVERAGE YIELD.

THE average yield of Indian corn in the United States for 1850 was, according to the census of that year, twenty-five bushels per acre ; the extreme limits being eleven bushels for South Carolina and forty bushels for Connecticut. For 1860 the census tables do not give the average product per acre for the whole country, nor do they furnish any returns from which the average yield for that year may be accurately determined. We have, however, numerous reports and estimates of acreable products from various sections of the country since that period, from which a proximate average may be arrived at.

Mr. Ezra Cornell has reported for Tompkins County in this State an average of 46.7 bushels per acre on the level of Cayuga Lake, and 32.4 bushels in localities one thousand feet higher. From other counties in the State there have been reports and estimates ranging from twenty-six bushels per acre to forty bushels and over ; making the probable average for New York between thirty-two and thirty-three bushels.

From Ohio we have returns, both official and otherwise, making the average product per acre in that

State, for a succession of recent years, nearly thirty-three bushels per acre.

In New England, acreable products have been estimated and reported from different States and sections, varying from twenty-seven to thirty-eight bushels, the most competent judges rating the average at about thirty-two bushels.

Some estimates from Indiana and Illinois would lead to the inference that the average for those States will reach from thirty-five to forty bushels per acre.

On the other hand, there are sections of the country of no small extent from which the reported estimates are lower than any of these figures. In some of the immense cornfields of the far West, and on the large plantations of the Southwest and South, the cultivation is necessarily imperfect and neglected, and the yield being correspondingly low, contributes to sink the average for the whole country.

Taking the various data and means of judging as we find them, though there is some room for difference of opinion, we may yet reach a general conclusion that can scarcely be very wide of the mark.

One writer puts the average yield for the whole country in 1860 at thirty bushels, another at twenty-eight and a fraction. The editor of the *Country Gentleman* places it in 1862 at thirty-five bushels. The opinion of the latter is entitled to great consideration; yet we are inclined to think that it is slightly above the mark. If we place the general average for the last five years at thirty-three bushels per acre, it cannot be very far from the truth.

The difference between the average yield of this grain and the amount raised per acre by many of the best farmers is at first view not a little surprising. When we observe scores of cultivators in every direction counting their annual yield by the hundred bushels per acre, and others ascending to still higher figures, and yet find that the average for the whole country during the past twenty years has ranged from twenty-five bushels to a little over thirty, we can scarcely credit or comprehend so strange a contrast. Yet the matter is very simple and easily solved. The difference in crops is a difference of diffused intelligence; and it is gratifying to know that the contrast is gradually melting away in the presence of farmers' clubs, and before the increasing circulation of farming journals.

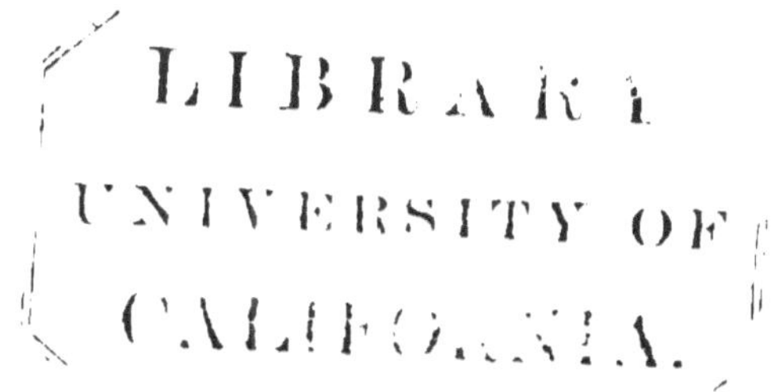

PRODUCTIVENESS.

THERE is no plant or vegetable grown by the farmer that is more variable in its yield, or more susceptible of the influences of soil, season, and treatment than this grain. Herein lies a strong argument for attending to its requirements, and studying out the conditions on which its productiveness depends. On the records of State and county fairs, and in agricultural and other journals, the crops frequently reported give striking proof of the prolific capacity of Indian corn, and well deserve the attention of the cultivator.

The following are a few of the large yields to be found on record, and may perhaps serve as a stimulus to our farmers, prompting them to aim at similar results. It should be remembered that large yields of corn tend to increase the supply of other provisions, and at the same time enable the farmer to keep up the quality of his land. Every man, therefore, who raises a large corn crop, not only improves his own condition, but contributes to the prosperity of his country.

David R. Bruce, of Desmoines County, Iowa, a lad of fourteen years of age, and L. H. C. Bruce of the same place, aged sixteen years, are reported in the *American Agriculturist* to have produced, the former one hundred and ten and a half bushels, and the latter one hundred and seventeen and a half bushels per acre without the aid of manure or fertilizers of any kind.

A writer in the *Country Gentleman* has stated that Joseph Wright, of Waterloo, N. Y., had not failed once in the previous three seasons to get over one hundred bushels of shelled corn to the acre, by planting the red-cob dent corn of Illinois, imported direct from the prairies.

The late Judge Buel, a most intelligent and enthusiastic cultivator, was an advocate of close planting in drills, in which he was successful, reaching from one hundred bushels to about one hundred and twenty bushels per acre. The Messrs. Pratt, of Madison County, by the same method succeeded in producing one hundred and seventy bushels to the acre.

The editor of the *Annual Register of Rural Affairs* states that one of the best farmers of his acquaintance has obtained one hundred and thirty bushels to the acre by planting his corn three feet apart each way.

The Browne corn has produced, as cited by Mr. D. J. Browne, in his *Memoir on Indian Corn*, one hundred and thirty-six bushels per acre, weighing fifty-eight pounds to the bushel.

The Whitman or Hill corn is stated by Mr. Fearing Burr, Jr., to have given a product of one hundred and forty bushels per acre.

It has been announced in a Kentucky journal that Major Williams, of Bourbon County, succeeded in raising one hundred and sixty bushels to the acre by planting in rows two feet asunder, with the stalks twelve inches apart in the row. This is another among many proofs that corn, if rightly treated, may be planted nearer than the usual practice without losing its earing capacity.

Mr. C. T. Johnson, of New Jersey, has reported to the Farmers' Club of the American Institute, a crop of the improved King Philip, reaching nearly two hundred bushels per acre, produced by close planting in drills.

In a field of corn of six acres, planted by Henry Norton, of Western Ohio, one-half the field receiving no manure, produced one hundred and twelve bushels per acre; while the other half, by subsoiling and liberal manuring, gave a product of one hundred and sixty-five bushels, the ears averaging nearly three-quarters of a pound in weight.

A. B. Miller, of Marion County, Iowa, has written to the *American Agriculturist* an account of several crops raised by farmers in that county in 1860, yielding from one hundred to one hundred and twenty-two bushels per acre; stating that another farmer in the same county, Mr. B. Long, has produced one hundred and seventy-eight bushels per acre on three contiguous acres; and still further, that Mr.

Long's son, under fourteen years of age, raised ninety-four bushels on half an acre.

A larger acreable product, however, than any of these, and probably the largest ever reached, was that of Dr. J. W. Parker, of Columbia, S. C. It is stated in the *Weekly Tribune* that the corn planted by him was the Bale Mountain, a variety obtained from North Carolina; that the land was under-drained, highly manured, highly cultivated, and closely planted, and that the yield was two hundred bushels and twelve quarts of shelled corn per acre.

But the prolific vigor of Indian corn is not limited to its yield of grain. The stalk crop is no less remarkable for its luxuriant growth and surprising product.

While the hay crop seldom exceeds two and a half tons per acre, averaging over the country probably not more than one and a half tons, the amount of stover accompanying the maize crop, forming a part of its product, and considered by many farmers quite equal in value to hay, generally ranges from two to three tons per acre, occasionally reaching four or five tons.

When the stalk crop is raised for the purpose of fodder exclusively, the yield is higher still. Nine tons of this fodder per acre, weighed after curing, are reported in the *Working Farmer* and stated to be sufficient in quantity for keeping ten cows seventy or more days. This amount has not unfrequently been equalled, and occasionally surpassed. In a report to an agricultural society of South Carolina, more than twenty-seven thousand pounds of cured stover are stated to have been produced on a single acre.

As a *green* fodder crop, raised for soiling cattle during summer and autumn, the weight of this stover per acre is still more remarkable. A writer in the *Country Gentleman*, over the signature of a "Buck's County Farmer," says that he has frequently raised from fifteen to twenty tons of green fodder per acre, and considers one acre sufficient in a good season for twenty head of cattle, from about the beginning of July to the middle of August.

Mr. John G. Webb, a dairy farmer near Utica, who usually plants ten or fifteen acres for summer feeding, reports his yield at twenty-five tons and upward per acre.*

R. H. Mack, of Parma, Ohio, in a communication to the *Country Gentleman*, gives twenty-two tons per acre as the result of his experience in growing stalks for soiling purposes.

S. W. Hall, of Elmira, N. Y., has raised thirty tons per acre by actual weight (as he states in the *Country Gentleman*), but considers this more than an average yield.

It has been stated in the *New York Daily Tribune*, that an acre has been known to supply over forty tons of green fodder; and a still larger product is given in *Allen's American Farm Book*, where one hundred and thirty-eight thousand eight hundred and sixteen pounds of green corn-stalks cut from one

* See "Tucker's Annual Register" for 1864, p. 99.

acre in a single season are quoted from a report to the Pedee Agricultural Society of South Carolina. This is the same crop which gave, *when cured*, twenty-seven thousand pounds, as quoted above.

It is not, however, to be inferred that such crops as the above are matters of course, or things of daily occurrence, nor that they are free from difficulty, or achieved without effort. The contingencies attending a large yield of corn are neither few nor trifling. But the persevering and resolute purpose of the well-informed cultivator is equal to them all, and the impunity with which his successful crop escapes casualties and defies contingencies, is an evidence how much can be accomplished when intelligence is guided by science, and industry is aided by skill.

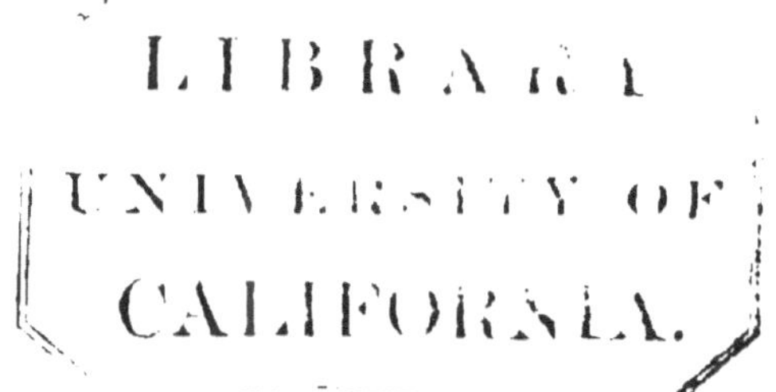

LIMIT OF PRODUCTION.

To the yield of this grain, as to that of every other, Nature has somewhere placed a limit, or rather, perhaps, has surrounded it with a series of limits, which no skill or ingenuity of man may exceed. There is, for example, a limit in the prolific power of the seed; another in the capacity of the soil; and still another in the area or space required by each grain for perfect development and fruition. These might be called, respectively, the limit of fecundity, the limit of fertility, and the limit of area, or distances.

It is safe to assume that neither of these has ever yet been reached. The productiveness of Indian corn has not yet been tested to its ultimate boundary. There is a possible yield greater than any yet accomplished. What that yield may be we do not know. It may be two hundred and fifty bushels per acre; probably more; possibly less. But what we do know is, that two hundred bushels per acre have been achieved. Beyond that lies the domain of uncertainty, a vast undefined region of dim twilight, which theory may explore, and experiment may develop, probably with useful results.

The prolific character of maize is shown, not more in the large crops spread over many acres, than in the self-multiplication of single grains. The reproductive vigor inherent in each separate seed is not a little remarkable. One kernel has been known to produce in a season several thousand grains, and single ears of the gourd-seed variety have produced more than a pint by measure.

Now, if the proximity of the growing grains did not interfere with this fecundity, if close planting interposed no limit to these prolific results, it is easy to see that an acre might be made to return a thousand bushels just as readily as it now returns a hundred. We know that a single stalk of maize will, under certain conditions, yield a pound or more of grain. And we also know that if an acre of good land, at the proper season, were literally covered with grains of corn, placed in contact and sprinkled over with earth, those grains, if all perfect, would all germinate. But would each one return a pound of corn? Certainly not; nor any other quantity. The close planting violates a law of Nature. There is a certain interval or space between the stalks that would render a pound of corn possible for each. There is another interval that would reduce this quantity to a gill; and still another that would render every stalk in the field grainless. These intervals, however, are not fixed quantities. They vary according to the soil, the kind of grain planted, etc. For each of these varying conditions there is some one mode of spacing better than any other—a certain arrangement of distances that will give a larger yield than any other. Let us suppose that yield to be two hundred and twenty-five bushels per acre. Then the spacing which gives that product is the best possible, and no deviation from those distances in planting would increase the yield. Here, then, would be a limit of production imposed by the law of distances.

But let us take another view of the matter. Every soil not absolutely sterile contains, in its natural state, a certain amount of the constituents of Indian corn. In a state of *perfect* fertility it would contain the *largest* possible amount of these, and in the exact condition and proportions required by the growing plants. We do not perhaps know what is the highest point of fruitfulness to which a given soil may be brought. But this is not material. The maximum of fertility is not indispensable for a maximum yield. If the space occupied by the roots of a single stalk contain one and a half ounces of the inorganic elements of corn, in the right condition and proportions, along with a small percentage of the organic constituents,* then such stalk should produce a pound of grain, so far as the yield depends on the prolific character of the soil; and if an acre of ground contain, in each square foot, one-half the above quantity of corn elements, then the capacity of such acre

---

* These being mainly derived from the atmosphere, and from descending rains, their presence in the soil is not required in the same proportions as the other class of elements.

is equal to over three hundred bushels of grain, so far as that capacity is determined by the fertility of the earth.

If, then, the farmer brings his land to this standard of fertility, complying at the same time with the other requisite conditions, he is entitled theoretically to expect a corresponding result. If he has made sure that his soil contains the constituents of maize in the ratio above given, he has reason to calculate on three hundred bushels per acre; and if he fails to get that amount, it is not the fault of the soil, but because there is another limit to the yield earlier reached than the limit of fertility. He is barred out by the limit of distances. If he had fertilized his soil to a capacity of five hundred bushels, yet by the hypothesis above stated, he could only get two hundred and twenty-five bushels, nor even that amount, unless he complied with the conditions on which it depends.

The only barrier, therefore, of any practical consequence to the farmer is that imposed by the law of distances. This limit, being the first that he reaches, renders any others that may lie beyond of little moment. He can raise but so many bushels on an acre as this principle permits; and how many that may be, experiment alone can determine. It is assumed above to be two hundred and twenty-five bushels, which is doubtless too low. It is extremely probable that the further improvement of existing varieties of corn, and modes of culture, and, still more, the introduction of new varieties, will yet prove that the real limit of production is in fact much higher.

3*

But the amount above stated may be confidently taken, for the present, as a possible yield, having been verified, on a small area of ground, in a number of instances. It is, in fact, probable that many farmers have produced, without being aware of it, even more than this, relatively, on limited portions of their fields.

Though it is, doubtless, true enough that results from small areas are not to be taken as certainties for large crops, yet it is also equally true, that experiments on a small scale are important and valuable for determining the best methods, and for proving, not indeed the certainties, but the possibilities for entire crops. The large yield obtained on one hundred square feet will not, of course, be so easily reached on an acre; yet the experiment, though small, will, if successful, be the sure precursor of a similar yield on a larger scale; for whatever is actually accomplished in the one case becomes undoubtedly possible in the other.

But after all that can be said, it must be admitted that the value of a large yield depends on what it costs to produce it. Nor is it at all likely that such a yield as the one above stated to be possible, would be found, *in the first instance*, a profitable crop. The processes by which it would be at first arrived at, would probably make it more than usually expensive. Still it would be a valuable result, and a point gained in the right direction. To reduce the cost of such a yield, would be a subsequent achievement, and one certain to follow, in due season. It is thus in a gradual way, and by single steps, that all valuable progress is made. It sometimes happens that these single operations, abstractly regarded, appear of little moment, and sink into temporary obscurity, till some thoughtful mind detects their importance as links in a valuable chain; and subsequent events ratifying the verdict, shed around them a halo of light in which the world discerns their true character.

VARIETIES.

THE varieties of maize are chiefly distinguished by—

1. The color.
2. The number of rows on the cob.
3. The size of the grain.
4. The form and hardness of the grain.
5. The chemical composition of the grain.
6. The color and size of the cob.
7. The length of time in maturing, etc.

From these and some other characteristics, and from their numerous combinations, have resulted an indefinite number of varieties, which have been still further increased by hybridizing and by change of climate. To repeat here the almost endless catalogue of existing varieties would be scarcely possible, and quite unnecessary. The following enumeration embraces most of the kinds in use, and all that are likely to be of any practical value to the farmer:

YELLOW CORN.

1. *New England Eight-rowed.*—This variety grows from six to eight feet high, with ears averaging nearly

ten inches in length, bearing a broad kernel of bright yellow. The number of rows is invariably eight, and the cob rather small. From this corn the King Philip and some other improved sorts have probably been derived.

2. *Golden Sioux*, or *Yellow Flint*, is a twelve-rowed variety, taking its name from the Sioux tribe of Indians, formerly resident in Canada, among whom it was first found. The grains are of medium size, and cob comparatively large. It abounds in oil, makes an excellent meal, and is very superior for fattening animals. It has been known to produce one hundred and thirty bushels to the acre.

3. *Canada Yellow.*—A small, early maturing, eight-rowed variety, with a small cob, and containing a large percentage of oil. It is much used for feeding to poultry, as well as to swine. It admits of close planting, and is quite prolific of ears.

4. *King Philip.*—An eight-rowed yellow or copper-colored corn, so called from the celebrated Indian chief of that name. It bears a long ear with a small cob, and the kernel is larger than that of the Golden Sioux. It is a hardy variety, ripening early, and very productive. It is much esteemed in New England, where it has been long cultivated, and is regarded by many as one of the best field sorts in use.

5. *Southern Big Yellow.*—This variety has a large cob, with the kernels large and very wide. It is partly of the nature of a Flint corn, but has less oil and more starch than the Northern Flint. It is late in maturing, but quite abundant in yield.

6. *Southern Small Yellow*, with grains similar in form to the preceding variety, but deeper in color. It matures earlier, is more oily, and less productive than the former.

7. *Dutton.*—This variety was introduced by Salmon Dutton, of Cavendish, Vermont. The stalk is of medium height, and the cob comparatively large, with ten to twelve rows of grain. The grains grow very compactly on the cob, and the ears being well filled out at the tips, and of a rich glossy color, present a very fine appearance. It is quite prolific, early maturing, and abounds in oil. It is capable of producing one hundred and twenty bushels to the acre.

8. *Browne.*—This is an eight-rowed sub-variety, improved from the King Philip by Mr. John Browne, of Long Island, in Lake Winnipiseogee. It has a small cob, with large grains, matures early, is very prolific, and being rich in oil is valuable for feeding. It admits of close planting, and has produced as high as one hundred and thirty-six bushels per acre.

9. *Rhode Island Premium.*—A hybrid variety of comparatively recent introduction, but quite popular in some parts of New England. It was produced by crossing the Canada, the Eight-rowed Yellow, and Red variety of Rhode Island. With close planting, it gives a very fair yield.

10. *Yellow Gourd-Seed.*—This is a cross of the Southern Big Yellow with the White Gourd-seed. It is a very prolific, many-rowed sort, with a small cob, comprising numerous sub-varieties, much in use

at the South and West. The ears grow very large, sometimes yielding a pound or more of grain.

WHITE CORN.

1. *Northern White Flint.*—This corn is semi-translucent, with a rather large cob. It is very similar in the shape of the ear to the Yellow Flint, and embraces numerous sub-varieties. The grains somewhat resemble those of the Tuscarora, but contain a large proportion of oil, and produces a substantial and excellent article of meal.

2. *Southern Big White,* with twelve rows of kernels, similar in form and size to those of the Big Yellow. It is a softer corn than the Northern Flint, containing less oil and more starch. It is consequently less adapted for feeding, and the meal is not easily kept sound for any length of time.

3. *Southern Little White.*—This has the grains smaller than those of the former, but similar to them in shape, growing more compactly on the cob, and containing a larger proportion of oil. This is not a prolific variety, and not extensively cultivated.

4. *Whitman* or *Hill.*—An eight-rowed variety, with a small cob, with the ears well filled out at the tips, and very productive. This corn is well adapted for feeding, but is not profitable for marketing, on account of the dull white color of the meal. It admits of close planting, and is a favorite kind in some parts of New England. It has been known to yield one hundred and forty bushels per acre.

5. *Tuscarora.*—This is an eight-rowed variety, with the kernel large, soft, and remarkably white. Though not a sweet corn, it is frequently used on the table in the green state. It is destitute of gluten and oil, and the meal when bolted resembles in appearance the flour of wheat.

6. *Long Island White.*—The ears of this variety are of good size, and usually contain from eight to ten rows. It is capable of a prolific yield, and produces a meal of sweet and pleasant flavor.

7. *White Gourd-Seed.*—In this corn the ears are shorter and much larger in circumference than those of the flint varieties, containing from sixteen to thirty-six rows of long, narrow kernels. It is a very prolific variety, extensively planted at the South, and is the source from whence many other sorts have been derived. Like other Southern kinds, it contains more starch, and less gluten and oil, than the flint corns of the North, and is therefore less suitable for shipping, and less profitable for feeding to fattening animals.

8. *Baden.*—This variety is an improvement of the White Gourd-seed, and takes its name from its founder. It is very productive, with a small cob, and grows to a remarkable size, yielding from four to six ears on a single stalk, and has been known to produce as many as ten.

SWEET CORN.

1. *Stowel's Evergreen.*—A late but prolific variety, with small cob, and long, deep kernels, which are much shrivelled when ripe. It is hardy, but tender,

continues long in a succulent condition, and is also an excellent variety to plant for soiling.

2. *Narraganset.*—A small early variety, with eight to ten rows and a red cob. It is sweet and tender, and very good to plant for a succession. It thrives best on a light soil.

3. *Rhode Island Asylum.*—The ears of this variety are large, with eight to ten rows. It is rather late, but productive, tender, and excellent in flavor. Its name is derived from the institution on the grounds of which it originated.

4. *Twelve-rowed Sweet.*—This is a late, hardy variety, with ten to fourteen rows. The ears are large, the yield certain, and the quality tender and excellent.

5. *Darling's Early.*—This is a sweet and tender variety, with eight rows, and of prolific yield. It may be planted for boiling until near the beginning of July.

6. *Burr's Improved Corn.*—A hardy and productive variety, with twelve to sixteen rows. The ears are of large circumference, and weigh, when fit for the table, from eighteen to twenty-two ounces. This corn is an improvement of the Twelve-rowed Sweet, and quite surpasses it in flavor.

There are many other valuable varieties of table corn, among which are—

7. *Adam's Early White.*

8. *Golden Sweet.*

9. *Mammoth Eight-rowed Sweet.*

10. *Mexican*, etc.

The foregoing enumeration embraces the leading varieties of field and garden corn. Besides these, may be mentioned the following:

Hæmatite, or *Blood Red*, of various hues, but more generally a deep red. It comprises a number of sub-varieties, some of which have a white, and others a red cob.

Rice Corn.—A small variety, so named from the resemblance of its kernels in size and form to the grains of rice. It abounds in oil, and is well calculated for feeding poultry.

Parching Corn.—A small variety, somewhat resembling the preceding. When parched, it is very crisp and tender, and of excellent flavor.

Chinese Tree Corn.—A variety in which the ears are suspended from the extremities of separate branches. An improved variety of this corn, which is said to yield seventy-five bushels per acre with ordinary culture, has been cultivated for some years by J. L. Husted, of Greenwich, Conn.

Oregon, or *Rocky Mountain.*—A peculiar variety, in which each kernel is enclosed in a separate envelope.

Egyptian Corn, with a head bearing some resemblance to millet.

IMPROVEMENT OF VARIETIES.

THE capability of improvement that belongs to Indian corn well deserves the attention of cultivators. Progress seems to be a law of its nature, and there is probably no variety at present known, however poor or however excellent, that may not be made better by adopting the appropriate means.

This progressive tendency is clearly seen on comparing the better sorts now in use with the primitive grain cultivated by the natives of this continent at the time of its discovery. The further we go back into antiquity, the fewer the sorts, and the poorer the quality appear to have been; and if the genealogy of this cereal could be traced to its source, it is extremely probable that all the existing varieties would be found to have sprung from one original stock, which was doubtless as much below the present standard as the untutored red man is inferior to the cultivated white.

The progress thus indicated in the past history of maize points clearly to an advancement in the future. The law impressed upon it at the start has never yet been suspended. Throughout animated nature the principle of life implies ceaseless activity and onward movement. To stand still is to stagnate, to deteriorate, and to decay. In obedience to this principle, no variety of Indian corn can long remain stationary. If neglected, it will degenerate. If rightly treated it will advance—slowly, perhaps, but surely, toward perfection.

The means by which this improvement is to be effected are extremely simple. So simple, indeed, that we might reasonably expect to witness greater progress than we have yet seen. In order to secure this object, the chief points requiring the attention of the farmer are *Selection* and *Culture*.

Every man who will exercise suitable care and judgment in the selection of his seed, without neglecting its subsequent cultivation, will find the quality of his grain and the amount of its product annually progressing; and the difference of a very few years will be so marked and unmistakable as to excite his surprise.

This principle of selection, if we did but realize it, is one of great extent and importance, and is capable of a very wide application. Its effects may be traced throughout the animal as well as vegetable kingdom, and the field of its influence is coextensive with the propagating universe. The valuable results it has accomplished, as seen in the various improved breeds of cattle, have long engaged the attention of farmers; and the practical application of the same law in the vegetable kingdom, though more recent, has been found no less favorable and certain in its effects.

"The principle of selection," says the editor of the *London Field*, "so successfully carried out among cattle and sheep, has of late been applied to the vegetable kingdom, and soon the various kinds of seeds bid fair to exhibit those qualities of superiority which can alone be produced by careful and continuous discrimination. . . . . In adopting selection, a great principle has thus been evolved, and one manifest advantage is that it is open to every agriculturist, without any additional expense to carry out the plan for himself."

Mr. Hallet, of Brighton, has applied this principle with great success to his wheat crop, and has been able by that means to more than double the size of the original ears. "It has been," he observes, "the great leading idea of my life, that the starting with an accidentally large ear is a very different thing from starting with a similar ear, the result of descent, or pedigree. Take the case of two heifers identical in every respect but pedigree—the one what she is by accident, the other by design. From the former you may get any imaginable kind of progeny, from the latter only a good kind. In other words, you have fixity of type; and the good qualities gain the force, as it were, of impetus by continual accumulation."

It is satisfactory to know that American farmers are neither indifferent nor inactive on this subject. Already marked improvements have been effected by this means in some of the varieties of Indian corn. The *Baden* variety, so named from its originator, is a striking illustration of this principle. It was produced from the White Gourd-seed, by Thomas N. Baden, of

Maryland, who, by a persevering and discriminating selection of the best seed for a series of years, with special reference to obtaining the greatest number of ears on a stalk, finally succeeded in establishing a variety which yields from five to seven ears, and which has been said to reach as high as ten ears to a single stalk. The *Browne* corn, an excellent variety obtained by improving the King Philip, is another illustration of this same principle.

NEW VARIETIES.

Closely allied to the improvement of maize by selection, is the introduction of new varieties by *crossing* or *hybridizing*. Here again the analogy drawn from the animal kingdom holds good, and the same law by which the better qualities of two different breeds of animals may be so blended in their joint off-spring as to form a third, different from either, renders it equally possible to combine the best properties of opposite sorts of maize into a new and distinct variety superior to both of its progenitors.

But here the principle of *selection* becomes more than ever important. This alone can give to the new hybrid that established character, or *fixity of type*, that shall render it reliable and of permanent value. "If nature be judiciously directed by art," said the late John Loraine, after a series of careful experiments, "such mixtures as are best suited for the purpose of farmers may be introduced in every climate in this country where corn is grown. And provided the desirable properties of any of the various corns be properly blended together, an annual selection of the seed, with care and time, will render them subject to very little injurious change. They do not mix minutely, like wine and water. On the contrary, like mixed breeds of animals, a large portion of the valuable properties of any one of them, or of the whole, may be communicated to one plant; while the inferior properties of one or the whole may be nearly grown out. When this object is obtained, and we become acquainted with the proper arrangement of the plants in our fields, so as to promote the utmost product, the crops of maize will by far exceed any estimate which would at this time be considered probable by those who have not carefully examined the economy of this plant."

To hybridize this cereal successfully does not require in the farmer any peculiar or unusual faculty; it is not the exclusive privilege of genius, nor the monopoly of gifted minds; but depends for success upon the plainer and more useful qualities of judgment, patience, and careful attention. A few leading principles are important to be observed, and those who may be inclined to undertake the propagation of new varieties, may perhaps find the following hints of some service :

1. Determine what precise traits or properties you intend the new corn to possess.

2. In selecting the sorts from which to propagate, prefer such as have these desired properties distinctly marked and predominating, with as few other prominent qualities as possible.

3. Let the varieties you employ be adapted to the climate.

4. Let the planting be so adjusted, as to time, that the tassels and silk fibres of all shall appear simultaneously. If these be not in unity of time, the hybrid effect will not be produced.

5. Every sample used to propagate from should be the purest of its sort, and if possible free from admixture. The more fixed and perfect the type of the several progenitors, the more certain and acccurately defined will be the qualities that mark the offspring.

6. All corn planted for propagating purposes should have every opportnnity of perfect development, by being placed in the best soil, at wide intervals, liberally manured, and well cultivated. It should also, of course, be entirely beyond the reach of the pollen of any other corn.

7. The surest mode of reaching the highest results in hybridizing, though it would require more time, would be as follows :

After carefully discriminating the several sorts to be used, let the cultivator improve each of these separately through a series of selections, as already explained, and then, by crossing, let him propagate the intended sort from the more perfect types thus obtained. The new variety resulting from this mode of proceeding would afterwards be kept pure and still further improved by continuing the same process of selection.

It would not perhaps be easy to foretell the extraordinary results that might and probably will yet be reached in thus improving and multiplying the varieties of Indian corn, by the joint aid of careful selection, judicious crossing, and thorough cultivation.

"This plant," says a writer in the *New York Daily Tribune*, "hybridizes with great facility. Some choice varieties have originated in this way, and others will undoubtedly be forthcoming, as no topic occupies more space in our agricultural journals than corn and its culture. Small fortunes have been realized by the originators of new strawberries, raspberries, and other perishable fruits. Others have grown rich by providing machines for shelling and grinding corn, and chopping the stalks into fodder. But to the fortunate author of a variety which will measurably supplant all others, there will be a rich reward."

We have every reason to believe that there is at least as wide a margin for improvement, in the case of Indian corn, as Webb and other eminent breeders have found, in the case of cattle and sheep. The results already achieved in this direction clearly enough indicate that a broad field for useful and remunerating effort is here presented to the cultivator.

Whoever will apply to this subject the requisite care, judgment, skill, and patience, will find ample compensation in the production of a quality of maize superior to any yet known. The competition is open to all. The humblest farmer in the country is just as likely, as the wealthy owner of a thousand acres, to be the founder of a new variety of corn that shall be, to all other varieties, what the South Down or the Merino is among sheep, or the stately Durham among cattle.

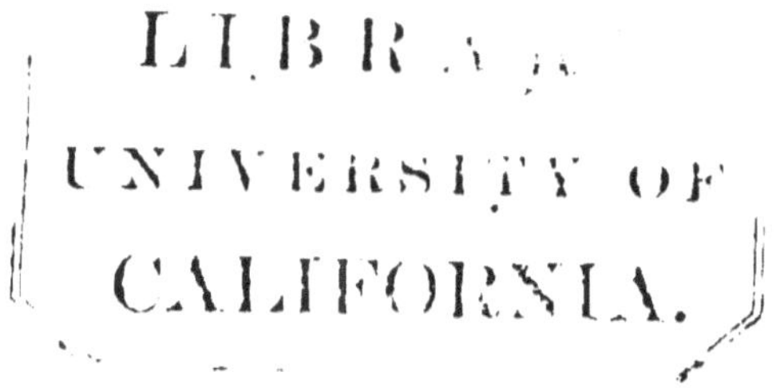

CHEMICAL ANALYSIS OF CORN.

THE chemical constituents of maize, according to Dr. Jackson, are starch, dextrine, gum or mucilage, sugar, gluten, albumen, oil, phosphoric acid, phosphate of lime, phosphate of magnesia, silica, potash, and oxide of iron. The proportions in which these elements are combined vary according to the variety of corn, and also, but in a less degree, according to soil and other circumstances.

A careful attention to the component parts of this plant, and a general acquaintance with the subject, are both useful and essential to the practical farmer. No man who goes on from year to year planting, cultivating, and harvesting his most important crop, without any definite idea of the elements composing it, can consider himself creditably posted in his business.

The following is the analysis of Dr. Dana:

Flesh forming principles, (gluten and albumen). 12.60
Fat forming principles, (gum, starch, sugar, woody
    fibre, oil, etc)............................... 77.09
Salts................................ .... 1.31
Water... .... 9.00
                                              ————
                                           100.00

In the ruta baga, according to Dr. Dana, the fat-forming principle amounts to 13 per cent., and in the potato to 24.34; while the proportion of flesh-forming substance in the former is equal to only 1 per cent., and in the latter to 2.07 per cent. As these roots are used, more or less, in feeding to stock, it is of some interest to the farmer to compare their nutritive and fattening properties, as here stated, with those of Indian corn:

ANALYSIS OF INDIAN CORN (when dried at 212° Fahr., to expel the water), by PROF. JOHNSTON.

Starch, etc.	71.6
Proteine compounds	12.3
Fatty matter	9.0
Husk	5.9
Mineral matter	1.2
	100.00

ANALYSIS OF PROF. PLAYFAIR.

Proteine	7.00
Fatty matter	5.00
Starch	76.00
Water	12.00
	100.00

The following table, by Prof. Johnston, gives the composition of the ash of *corn-stalks*, as compared with a similar analysis of the straw of wheat, barley, oats, and rye. The proportion of each constituent is given for one thousand pounds of the ash:

	Corn Stalks.	Wheat Straw.	Barley Straw.	Oat Straw.	Rye Straw.
Potash	96	125	92	191	173
Soda	286	2	3	97	3
Lime	83	67	85	81	90
Magnesia	66	39	50	38	24
Oxide of Iron	8	13	10	18	14
Phosphoric Acid	171	31	31	26	38
Sulphuric Acid	7	58	10	33	8
Chlorine	15	11	6	32	5
Silica	270	654	676	484	645
	1,012	1,000	963	1,000	1,000

The ash of the *grain* of each of the above, when analyzed, gives the following proportions:

	Corn.	Wheat.	Barley.	Oats.	Rye.
Potash		237	136	262	220
Soda	} 325	91	81		116
Lime	14	28	26	60	49
Magnesia	162	120	75	100	103
Oxide of Iron	3	7	15	4	13
Phosphoric Acid	449	500	390	438	495
Sulphuric Acid	28	3	1	105	9
Silica	14	12	273	27	4
Chlorine	2		Trace.	3	
	997	998	997	999	1,009

These tables will serve to guide the farmer in the
application of fertilizers to his corn. They indicate
the proportions in which the various constituents of
both the grain and the stalk should be found in the
soil. If, for example, he is about to plant a corn crop
exclusively for the fodder, he finds that soda and
silica are required in the soil, in far larger proportions
than any other inorganic element, and next to these

phosphoric acid. If, on the other hand, his corn is planted primarily and chiefly for the grain, he learns that phosphoric acid is required in a proportion nearly equal to that of all the other elements together, and that next to this in importance are potash and soda.

An inspection of these tables will also throw some light upon the relative feeding values of corn-stalks, and the straw of the other included grains, as well as upon the comparative nutritive values of the grains themselves.

The proportion of ash contained in any plant or grain represents the amount of inorganic matter that enters into its composition. When the plant is burned, all the other constituents, amounting generally to over ninety per cent. of the entire weight, disappear. We are thus able to determine what grains contain the smallest proportion of inorganic matter, and are consequently least exhausting to the mineral elements of the soil.

In the following table, Prof. Johnston has given the quantity of ash yielded by one thousand pounds of each of the plants named:

Indian Corn	15 lbs.	Corn-stalks	50 lbs.
Wheat	20 "	Wheat straw	50 "
Barley	30 "	Barley "	50 "
Oats	40 "	Oat "	60 "
Rye	20 "	Rye "	40 "
Peas	30 "	Pea "	50 "

The investigations of Dr. Jackson, of Boston, in regard to the properties of corn, are equally curious and instructive. Among other interesting facts, he has shown that the proportion of phosphates in each variety of maize depends on its assimilating power. It was found that of two varieties of corn (Tuscarora and sweet) *growing on the same cob,* the former had less than half the amount of phosphates contained in the latter.

To those who have not seen the report of Dr. Jackson, a brief statement of his further researches will perhaps be interesting.

In most of the yellow varieties, the oil is the seat of color, the hull or epidermis being transparent. In the white varieties, the oil being colorless and pellucid, and the hull transparent, the farinaceous portion of the kernel, which is white, gives a similar appearance to the grain. In the hæmatite varieties the red, purple, and blue colors are chiefly derived from the epidermis.

The proportions of oil vary from six to eleven per cent.; the flint corns of the North being found to contain more than the Southern varieties. The oil is analogous to animal fat, and is readily converted into that substance by a slight change of composition.

The gluten and mucilage contain nitrogen, which is necessary to the formation of fibrous tissue, muscle, nervous matter, and brain.

Starch is convertible also into fat and into the carbonaceous substances of the body, and during its slow combustion in the circulation, gives out a portion of the heat of animal bodies; while, in its altered state, it goes to form a part of the living frame. Sugar acts in a similar manner as a compound of carbon, hydrogen, and oxygen, in the formation of fat of animal bodies.

From the phosphates the substance of the bones and the saline matter of the brains, nerves, and other solid and fluid parts of the body are in a great measure derived.

The salts of iron go to the blood, and constitute an essential portion of it, whereby it is enabled by its changing degrees of oxidation, during its passage through the lungs, arteries, and veins, to convey oxygen to every part of the body.

Thus it appears that in each kernel of corn all the elements have been deposited by Nature, that are essential to a healthful, invigorating, and nutritious food.

DEVELOPMENT AND STRUCTURE.

THE vital principle of maize is lodged in the embryo, or rudiment, a small, clearly defined interior division of the seed, or kernel. This embryo is the starting point of life and growth. It extends from the base of the grain upward, about two-thirds of the distance toward the crown, and lies in contact with the epidermis on one side of the kernel, through which it can be distinctly traced by the eye.

The earliest movement of the seed in developing the new plant is termed *germination*. When the plant has advanced so as to form leaves that contribute to its growth, the process is termed *vegetation*.

Three conditions are essential before germination can take place. The presence of heat,* moisture, and air is indispensable. After the seed is planted, and these agents have had time to exert their quickening influence, a small root shoots out, with a very rapid growth, from the base of the embryo, and, after another interval, the stem rises slowly from its apex.

* *48° Fahr. is about the limit of temperature, below which corn will not germinate.*

The progress made by the roots during the first few days is quite remarkable. They not unfrequently attain to a length of fifteen or eighteen inches before the stem has made three inches above the surface of the ground.

From the relative positions of the stem and the early roots, the former springing from the crown, and the latter from the base of the embryo, it is evident that the most natural and favorable position of the grain for incipient growth is with the base downward and the crown above. When this condition is reversed, as continually occurs in planting, the stem and root are each compelled to describe a curve, sometimes equal to a half circle, in order to acquire their normal position. When this position is reached, if the seed should be turned over, the stem and root would again promptly bend themselves through another curve, to recover once more the situation natural and indispensable to their proper growth.

That the position of the kernel when planted is calculated to affect the progress of germination is an obvious and natural conclusion. The author has found, in some experiments having reference to this point, that grains planted in an inverted position are retarded from ten to fifteen hours in the time of their appearance above ground, as compared with others planted in an upright position.

As soon as the germination of the seed begins, the stem, obeying a natural instinct, springs upward toward the sunlight, while the roots, equally obedient to an instinct of their nature, travel downward into the earth, and away from each other, spreading themselves in every direction, and penetrating many thousand cubic inches of soil, in quest of nutriment to satiate a voracious appetite that began with their existence, and will only be extinguished at their death.

The natural proclivity of the roots of plants to push their way into congenial darkness, and of the stem to seek the presence of the light, may be illustrated by a simple experiment. One, among several tried by the writer, for the purpose of observing the early tendencies of germination, gave a very clear result. Having planted some grains of maize in glass jars filled with earth, the kernels being arranged against the side of the glass, one of these jars was placed in a dark room, and the other exposed to the light of a window.

After an interval of about thirty-six hours the roots began to show themselves, and after another brief interval the stems made their appearance. The only peculiarity about the latter was, that in the jar exposed to the light, they assumed, before reaching the surface of the soil, the green tint peculiar to the stalk and leaf above ground, while in the other, they remained nearly white after rising above the soil. In the jar from which the light had been excluded, the roots formed rapidly and abundantly against the side of the glass, while in the other jar they retreated from the glass almost in a direct line, evidently shunning the light, and seeking to hide themselves in the recesses of the soil.

When in the progress of its growth, the stem of

the corn plant has struggled up from its earthy bed, and approaches the point where germination ceases and vegetation begins, it pushes its bodkin-shaped cylinder of compact foliage through the surface of the earth, changing its color at once from white to green, and opening out its uppermost leaves to enter upon their function of respiration.

As the growth advances, other rolled-up leaves are successively developed from the crown of the stalk, until the tassel is fully formed and the plant assumes its perfect outline. The leaves grow broader and longer as they rise, one above the other, from the base of the stalk more than half way to the summit; after which they gradually and uniformly diminish in size to the uppermost leaf which is near the tassel. " One leaf grows from every joint in the stalk, but in such a way as to alternate sides. The first formed leaf, and after this every leaf in regular succession, clasps the stalk closely until it approaches near to the under side of the leaf above; after this it grows out from the stalk, and a beautiful fan-like appearance is at length produced which is not equalled by any other annual plant cultivated for the value of its fruit."—*Farmer's Encyclopædia.*

The stems on which the ears are formed proceed from the joints, commencing usually at the one nearest the ground. The number of ears on a stalk vary from one or two, to five or six, in rare cases reaching as high as seven or eight; though it is not often that more than two or three ears are matured on the same stalk. The ranks of grain on the ear vary in number from eight to thirty-six, being always an even number, and the product of single ears is about five ounces on a general average, though occasionally reaching over a pound.   The dimensions of the ear range, according to the variety of grain, from less than two inches in length in some of the dwarf varieties, to over sixteen inches in the largest, and sometimes reaching, in the gourd-seed variety, more than half that number of inches in circumference.

From the extremity of each ear flows out a cluster of soft and silk-like fibres falling like drapery over the husks.   These little threads are charged with one of the most important functions in the whole economy of the plant.   Each fibre proceeds from a separate grain, and every grain on the ear has a fibre to represent it.   The *Farina fecundans* is a fine, light, powdery substance dislodged by the wind from the flowering tassel that crowns the stalk.   This powder or pollen, descending from the tassel, lights upon the silken drapery of the ear, and the rudimental grains are thereby fertilized.   In the absence of either fibre or pollen, or even in the failure of their contact, the result would be, not an ear of corn, but a naked cob.

How curious and inscrutable is this recondite process !   How full of mystery indeed are all the processes of vegetation ; and how humiliating to the towering faculties of man to reflect, that though his mind may range at will through infinite space, measuring spheres and orbits and periods of revolution with amazing accuracy, penetrating sidereal systems on the confines of creation, and aspiring to embrace the universe in its grasp ; yet when he walks abroad in the vegetable kingdom of his own little planet, at every footfall he treads upon a mystery, and on every side his intellect is overmatched by each tiny flower and every blade of corn !

The wide range of climate in which Indian corn can be grown to maturity necessarily occasions a marked difference in the length of its season, or the time it requires for ripening. This period varies from two months to six or seven; and some precocious kinds, in high latitudes, are found to ripen in less than sixty days.

The average rate of daily increase in the size of the stalk, during the period of growth, differs with the climate, the soil, and the variety of grain. In some observations made by the author, the growth was found to be seventy inches in fifty days, being an average of one and four-tenth inches per day. The greatest increase noticed in a single week was twenty-two inches, and in a single day four and a half inches. Some of the largest varieties, especially in warmer latitudes, would probably show a more rapid growth than this.

But an increase of even four inches in twenty-four hours, though small when compared with some other instances of vegetable growth, is yet, in one aspect, curious and remarkable. The movement of this increase, which is equal to an inch in six hours, slow as it seems comparatively, may be converted, under a powerful lens, into a velocity of two inches, or more, per minute—a rate of motion easily detected by the eye.

In thus bringing the *movement* of vegetable growth

under the distinct perception of one of the senses, the mind seems to come into closer contact with the mysteries of vegetable life.

The height to which this cereal is capable of attaining is exceedingly variable. It is determined in part by the soil, in some degree by the climate, but depends still more upon the variety of grain. It ranges from less than two feet to over fifteen, and in tropical climates a still larger and ranker growth is not unusual.

The roots, in a deep, mellow, and fertile soil, are capable of penetrating to a depth of over two and a half feet, and horizontally have been traced to a length nearly equalling the height of the stalk. The prop-roots appear at that stage of the growth when the increasing size and weight of the stalk, and the accession of tassel and ears, render such support needful. They usually spring from the first joint above the ground, taking an oblique direction toward the earth, which they soon reach and penetrate, spreading through it in search of nutriment, and anchoring the stalk more securely to the soil.

The juices that nourish the plant are absorbed from the earth through the fine and thread-like fibres of the roots, passing in succession through the roots of large size until they reach the stalk, from which they are transmitted to every portion, and to the smallest extremities of the plant. From the leaf-stalk this sap is distributed in very minute veins through the whole expanse of the leaf, which brings it in contact with light and air. The watery portion of the sap is here in part exhaled, while carbon and oxygen are alternately imbibed and given off. "In the day-time," says Professor Johnston, " whether in the sun-shine or in the shade, the green leaves are continually absorbing carbonic acid from the air, and giving off oxygen gas. . . . . . When night comes, this process is reversed, and they begin to absorb oxygen and to give off carbonic acid. But the latter process does not go on so rapidly as the former ; so that, on the whole, plants, when growing, gain a large portion of carbon from the air." Thus does respiration keep up its unceasing work through the leaves or lungs, and, by appropriating from the air with nice discrimina-tion precisely what the plant requires, and rejecting whatever is needless or hurtful, purify it from noxious elements, and minister to its healthful growth.

In whatever light, then, we contemplate this inter-esting plant, whether in its curious structure, or in the processes of its rapid and vigorous growth, or in the flowing and graceful outlines of its foliage, or in its tall, erect, and majestic stature, we equally recognize the hand of its Author, who has attested its value to man, by impressing upon it the stamp of nobility and clothing it in forms of beauty.

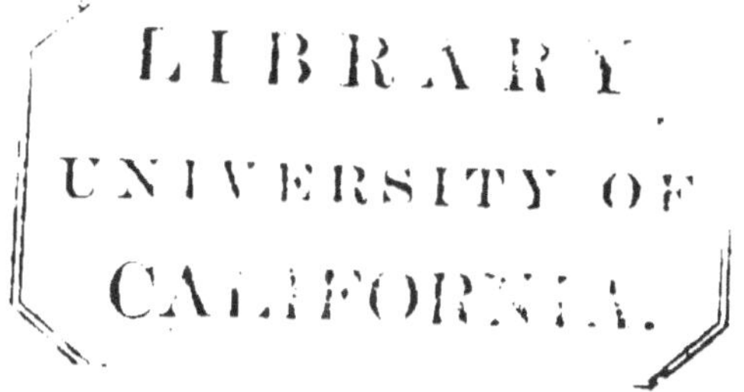

SEED.

I. SELECTION OF SEED FOR PLANTING.—That the quality of the seed planted by the farmer has a material influence on the quality and amount of the resulting crop is a matter that every practical man well understands. The importance, therefore, of giving the most careful attention to the selection of the seed is perfectly obvious. No man who neglects this essential point can place any reliance upon his crop. If his seed-corn is not properly sorted out, he cannot be certain of its kind, its value, or its results. If he does not know what he plants, how can he be expected to know what he is going to reap? His crop will be a lottery, with more blanks than prizes, and he can form no reasonable calculation in regard to it, either as to quality, certainty, or amount.

On the other hand, the man who in due season gives thoughtful heed to the selection of his seed, spending an ungrudged hour in his cornfield at the right time to secure the most perfect ears of grain, as the germ of a future crop, will be morally certain of at least a reasonable success. He has made a good beginning for another season. The first step is well taken and in the right direction.

The following rules for the selection of seed-corn, suggested by the experience of practical cultivators, will perhaps be of service to the farmer as a guide in making his selection :

RULES FOR THE SELECTION OF SEED-CORN.

1. The most essential point to start with is a *good variety*. No correct farmer will plant or use on his farm any but the best grain. If, therefore, the corn you have been raising is an inferior kind, abandon it at once, and procure the best variety that will succeed in your locality. Begin with the purest and most perfect seed you can obtain, and you will easily be able to keep it pure, and make it continually better by attending to these rules.

2. Select your seed from those stalks that have the most ears, taking the best from each stalk.

3. The earliest ripe in the field is to be preferred, unless otherwise objectionable.

4. Those stalks that bear their ears nearest the ground are the best to choose from, provided the ears are right.

5. Select large, fair ears, with kernels of a bright, clear color.

6. Prefer those ears in which the rows are most regular, and the grain most uniform in size.

7. Choose those ears that taper the least, having their butts very little larger than their tips.

8. Of several ears on the same stalk, those that grow nearest the ground are to be preferred, if they have the other requisite points.

9. Select such ears as grow upon the shortest footstalk.

10. Those ears that are well filled out at the tips, with the grain covering the extreme end of the cob, are much to be preferred.

11. From each ear take the central grains, rejecting tips and butts. It has been satisfactorily proved that the kernels near the ends of the cob give a smaller yield and an inferior grain.

12. If you plant seed not raised in your own vicinity, let it be from a colder rather than a warmer region.

13. It is an excellent plan to appropriate a small piece of ground for raising seed-corn, at a distance from the main crop. In doing this, select a warm situation, free from excessive moisture, and let the ground be subsoiled or trenched, thoroughly pulverized, and well manured. Plant in hills four or five feet apart each way, with six to eight grains in a hill, thinning out afterwards to two or three stalks. The advantage of planting more than you intend to leave is not merely that it provides for worms and accidents, but it gives a chance for preference or selection. When the corn is up eight or ten inches, you will often find a material difference between the best and poorest stalks. You thus have an opportunity of selecting the best. The greater the number you have to choose from, the greater is the chance for perfection in those

selected. Your seed-corn being now well planted and fairly started, with proper attention and care in the further management of it, you cannot fail to secure a fair proportion of large and beautiful ears of perfect grain.

By following up this system, the farmer will discover, at the end of a very few years, that his corn has gained many fold in yield, and still more in quality. The advantage attending a discriminating selection of seed is well established by the uniform results of experience, and it seems incredible that any cultivator can be indifferent to a matter of so much consequence. He may bestow any amount of labor upon the tillage of his field, and any amount of expense upon the manure, yet if he plants an inferior grain, he can only gather an inferior crop. The difference between thirty or forty bushels per acre, and sixty or seventy bushels, may very possibly prove to be, in practice, a mere question of seed. Whether his crop will return him ten per cent. or fifty per cent. on the cost of it, may depend entirely upon the single hour that he did or did not employ in selecting his grain for planting. If such considerations as these, that go right into the farmer's pocket, are not sufficient to arrest his attention and influence his practice, his indifference may indeed be considered hopelessly incurable.

II. PREPARATION OF SEED FOR PLANTING.—It is a very general practice, with the best farmers, to steep the seed of this grain before planting, and the practice seems to be justified by reason and experience.

It is attended with a twofold advantage: in quickening and promoting germination, and in offering a means of protection against the earliest and most dangerous enemies. There are various liquid preparations employed for this purpose. Some of the more usual are solutions of saltpetre, guano, copperas, wood ashes, etc. The sulphate, nitrate, and muriate of ammonia, and chloride of lime have also been used with advantage, as well as urine, and other forms of liquid manure. These solutions, however, require to be used with caution, and most of them should be made very dilute.

Some cultivators are in the habit of employing powerful solutions, and others recommend to raise them to a very unusual temperature, as if they imagined that some extraordinary effort in starting the crop were going to have the effect of a charm all the way through. But the object of steeping is to promote, not merely a quick but a *healthy* germination; and this is not to be accomplished by the use of *excessive* stimulants. A morbid growth, however rapid, is no ultimate advantage. The results of experience combine to prove that in this, as in every other stage of the growth of corn, there is nothing gained by doing violence to the processes of Nature.

Some solutions are more effectual than others in protecting the grain against its enemies. Saltpetre and copperas are each considered good for this purpose, but a moderate coating of tar is found to be still better, and this practice is now pretty generally adopted.

The late Judge Buel recommended a moderate solution of crude saltpetre, to which he added half a pint of tar for eight quarts of seed; the tar previously diluted with a quart of warm water. The mass is to be well stirred, and when the corn is taken out, let as much plaster be added as will adhere to the grain. The experience of years, he adds, will warrant me in confidently recommending this as a protection for the seed.

Coal or gas tar is now preferred by many farmers, and when used should be limited in quantity and applied as evenly as possible. Mr. G. Haines, of New Jersey, in writing to the *Country Gentleman*, remarks: "I have used both kinds of tar for that purpose, but for the last ten years have preferred gas or coal tar, because it is much more easily applied, and equally safe. If the corn is made jet black with it, it may not grow, but there is no occasion for that. Take a paddle and dip from the tar to the corn once or twice, then stir till the corn is all coated, and appears through the tar of a yellowish brown color. It may easily be tested by throwing a little to the poultry. The crow blackbirds have about twenty nests in the pine and cedars of my yard each spring; but if my planted corn was tarred (which is generally the case), I have not the slightest objection to it."

Mr. G. F. Saxton, of Williston, Vt., writes to the American Institute Farmers' Club as follows: "You are mistaken in supposing coal tar will injure seed corn. I have used it for five years upon seed for several acres annually with perfect success, as follows:

Soak the seed ten or twelve hours, drain off the water, apply the tar immediately in proportions of half a pint of tar to one bushel of corn, and stir until coated equally. If the corn is cold it is better to put hot water with the tar to thin it, as much water as tar, as it will be easier mixing. If this mode is followed, I will warrant the seed to grow as well as without tar."

In the further discussion by the Club, it was re-marked: "We are glad to be set right by a practical man in relation to the use of coal tar. We will also state in this connection, that it is recommended as a good preventive of the ravages of worms and bugs.

"Adrian Bergen said he always soaked and tarred his corn, and believes the tar some protection against crows as well as insects.

"John G. Bergen said the trouble about using coal tar is that those who have complained of its in-juring the seed have used too much. The quantity recommended by Mr. Saxton is quite sufficient for the purpose for which it is applied, yet not enough to in-jure the germ. To obviate the trouble of seed stick-ing to the hands, mix it with dry ashes, plaster, or dust."

TIME TO PLANT.

The proper time to plant corn depends on circumstances so many and various, that no specific rule can be laid down on the subject. It differs according to the variety of grain planted, the character of the soil, the climate, the season, etc. Between the extreme northern and southern sections of the country, the difference of time amounts to three or four months. In some parts of Maine and Minnesota the usual season for planting is June; while in Florida or Louisiana it is usually March. Throughout the Middle States and most of New England, the period considered safest, as a general rule, is the middle of May. Yet such is the difference of seasons, that in some years a crop planted during the last week in April, and in other years the first week in June, would give a better result than if planted at the middle of May, showing a difference of more than a. month *in the same latitude*, produced by a difference of seasons.

Thus it appears that the vicissitudes of the weather in different years have a more disturbing effect on the time for planting than any of the other causes. In fact this question of fluctuating weather, of early or late season, is after all the only real difficulty in the case, and the one on which all the others depend. The other contingencies are made so by this. They are variable, but all of them are determinate. If, therefore, the question of soil, of latitude, and all the other variable elements could be separated from the vicissitudes of temperature, the time for planting corn, so far as relates to them, might be reduced to fixed rules.

It is true that latitudes vary, and each different degree requires a different period for planting. Yet every farmer knows that his latitude is a fixed, assignable figure, and that it always remains the same. It differs from that of other men, but for him it is unchanging. The same is true in regard to soils. A sandy loam may require a period for planting different from that which would suit a tenacious clay. But the farmer who has a sandy loam one year, will not find it changed into clay the year following. Though soils differ for different individuals, yet for each man they remain the same. So also in regard to all the other circumstances affecting the question.

Could we, then, reduce the inconstancy of the weather to a condition of like certainty, or bring it within determinate limits, it would be quite possible to assign a precise day of the month for each kind of soil, for every variety of corn, and for every degree of latitude, which might be adopted in planting with perfect safety. We might lay down an accurate time-table for planting corn that would apply to the whole country, and meet the case of every farmer.

But, unfortunately, the question of season is not determinate. Temperature rises and falls according to no settled or ascertained law. Frost comes and goes apparently at the dictate of its own humor; and the weather is capricious to a proverb, and filled with elements of uncertainty. Man has learned to explore the earth, and detect the causes of its fertility, to regulate its production, and make it obedient to his purpose. But he cannot subdue the atmosphere to his will, nor assign limits to its phenomena. He can classify all the plants in the vegetable kingdom, and tell with accuracy their times and seasons; but he cannot reduce storm and sunshine to a system, nor bring the clouds up to time. He may subdue the most incorrigible soil, but he cannot subjugate the thermometer. He can dominate the mysterious energy of the electric fluid, compelling it to traverse the bed of the ocean, or to circulate around the globe on aërial wires to give swift wings to his flashing thought; yet can he not arrest for a single hour, nor even predict, the fall of the mercury that shall blast a thousand crops.

Thus science becomes the sport, and man the victim of fluctuating weather. Subject to no fixed laws, and recognizing no assignable limits, it defies alike all human calculation and human control. It comes into the arrangements of husbandry with the reckless power of an autocrat, setting aside appointed days, and thwarting plans innumerable.

On this subject, therefore, the farmer is left to depend very much on his own resources. Yet in all this he finds no occasion for despondency. He finds that a sound judgment carefully exercised in the light of the experience of former years, and guided by those hints and indications that Nature is ever presenting to inquisitive minds, will nearly always shape out for him the course of safety and success. In settling practically the question when to plant his corn, he banishes from his mind all those maxims that embody their entire wisdom in a specified date, or in a prescribed stage of the moon, and examines the condition of the soil and the state of the vegetable world for traces and indications more to be relied on.

"There is a *right* and a *best* time for planting corn," says a very sensible writer in the *Country Gentleman*, "and by employing *just that* time for the purpose, a farmer may all the more confidently calculate, if he do not fail or err somewhere else, on raising a *maximum* crop, not only of the grain but of the stalks also. And the right and best time is to be discovered, not by the almanac, nor by the practice of neighbors, who 'think that from the 10th to the 20th of May is the proper time for planting,' nor by blindly copying after some one whose whim it may be to plant 'seldom or never later than the fifth day' of May, but simply by observing the progress of vegetation in soils resembling that in which the planting is to be done. Vegetation will start sooner in sandy loams, and all such soils as contain much sand or humus, than in those in which clay predominates.

Making allowance for this fact, the right and best time for planting corn, let the latitude and the locality be what it may, is to be discovered and determined by observing the natural vegetation. Whenever there is *good* reason to think that the ground is warm enough to cause a speedy germination and growth, then is the time to plant. And to ascertain this, I know of no rule so safe and sure as that which Judge Buel taught me and others many years ago, namely, to plant when the apple is bursting its blossom buds."

But human judgment is not infallible; and if the husbandman is not always sure of his time in planting; if, notwithstanding the utmost care and attention, he discovers that his grain has been committed to the earth a little too soon, or a little too late, he yet finds with satisfaction that the consequences are not very serious, if he has faithfully pursued the right methods in planting, and in the treatment of the soil. The careful, well-informed farmer, the man, who, by reading, adds the experience of others to his own, has always a twofold advantage in such cases; for he is not only less likely than others to commit an error, but in case an error should be committed, he is measurably insured against the consequences by the resources of skill and science which have already been employed in his favor, and which are still at his command for any emergency that may arise.

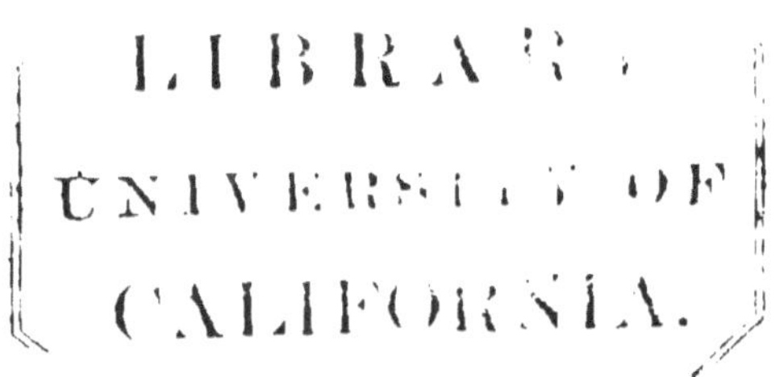

THE SOIL AND ITS CONSTITUENTS.

Although Indian corn will grow, as already stated, on nearly every kind of soil, from the lightest sand to the heaviest clay, yet, like other plants and grains, it has its preferences, and the interest of the farmer lies in consulting these as far as possible. However well it may succeed on lands where other grains would fail entirely, or make a feeble growth, it is only in a well-adapted soil that its best capability is developed. Give it a congenial element, in which its hungry roots can range and riot without limit, and it will make generous returns, that will even exceed the liberality of the treatment. "It delights," says Mr. Harris, "in a loose, pliable, warm, porous, deep soil, abounding in organic matter. It does well on all good wheat soils, yet it often does better on soils too light and mucky for wheat. It is a gross feeder. We can easily make land too rich for wheat, but I have never yet seen any too rich for the production of Indian corn."

The fertility of soils is determined chiefly by the amount of available plant-food contained in them. The cereals, and nearly all cultivated plants, are found to contain more or less potash, soda, lime, magnesia, silica, alumina, oxide of iron, oxide of manganese, sulphuric acid, phosphoric acid, and chlorine. There are three other substances, iodine, bromine, and fluorine, that enter into the composition of most plants, but in proportions so minute as to be of no practical importance. The first-named substances, eleven in number, constitute the inorganic parts of a plant, or that portion which it derives entirely and exclusively from the soil. Hence these elements, in one proportion or another, will be found contained in every well-conditioned soil.

There is evidently, therefore, in corn-culture, but one proper course for the farmer to pursue. It devolves upon him to ascertain, as nearly as possible, what proportion of the constituents of maize his soil already contains, and in what condition these constituents exist. The latter point is especially important; for whatever be the quantity of them, unless they are in such a state that the plant can appropriate them, they might nearly as well be entirely absent.

On this subject the science of chemistry will enlighten the farmer up to a certain point; beyond that he must rely upon other sources of information. Chemical investigation will determine, with sufficient accuracy, the elements of the soil on the one hand, and the elements of maize on the other; and a comparison of these would seem to indicate precisely what ingredients are yet wanting for the intended crop. But this indication is, after all, not entirely reliable. As the constituents of plants exist in the soil in various conditions, it is necessary to know, not merely whether they are present, but whether, also, they are in that peculiar state in which the growing plant can use or appropriate them. This condition chemistry has not yet been able to discriminate with certainty. It may, indeed, determine very correctly what proportion of potash, or soda, or phosphoric acid is contained in a cubic foot of any given soil; but what the cultivator needs to know is, how much of these substances it contains in that state, that will enable them to minister to the *immediate* wants of the plant.

Nearly all soils contain, in a state of nature (as elsewhere remarked), the principal elements of maize, in greater or less quantities, and some of these elements are found in proportions even much larger than the plant requires; but their value depends entirely upon their state of adaptation. If, from their peculiar combinations or other causes, they are impervious to the descending rains, and unfitted to the requirements of vegetation, they add nothing to the present fertility of the soil. There are fertilizing elements in the hard impracticable rock, and the chemist can doubtless determine the proportions of them; but it does not follow that the rock or any part of it is *at present* an available soil for the growth of plants. The analysis that reveals the relative quantities of plant-elements, leaves the quality and fitness of them still obscure and uncertain.

If the chemist could indeed resolve the soil into its elements, with an absolute precision and certainty as to the *condition* of each ; if, while he tells the farmer exactly what proportion of each constituent of corn is lodged in every square foot of soil, he could also tell him, with the same accuracy and certainty, *what part of that proportion is perfectly adapted to the immediate use of the growing plant,* the effect would be most remarkable. Fertilization would be reduced to an exact science, and agriculture would be revolutionized.

But though chemistry, that has done so much for agriculture and for the other useful arts, has not yet achieved this needed revelation, it has before it nevertheless, like other sciences, a future of indefinite possibilities; and there is some reason to believe that the time will arrive when the analysis of the soil will be so thorough and complete as to disclose to the cultivator not only this information, but all else in this connection that he needs to know.

But meantime the question remains, How is the farmer, while waiting for this chemical illumination, to obtain the desired information ? How is he to know what amount his soil contains, and what amount it lacks of the *available* elements of his grain ?

The answer to this inquiry is plain and simple. There is just one method, and only one, of arriving at this knowledge—consult Nature. Interrogate the soil in a series of experiments. This is an old doctrine, but a very sound one, and no less true to-day than it was in the time of Lord Bacon. The testimony of Nature can always be had, and is always more valuable

5*

than any other. Put your soil on the witness-stand. Subject it first to an examination direct, and then to a rigorous cross-examination, and you will compel it to disclose those reluctant secrets that chemistry has not yet arrived at.

PRACTICAL MODE OF TESTING THE SOIL.

In order to determine what manures are best adapted to a given soil, there is no method more certain and successful than to institute a series of trials or experiments, which, if well devised and rightly conducted, will enable the farmer to understand the wants of his land, so as to proceed intelligently in supplying them. These trials may be, for the most part, accomplished in one season, but require for the best and the most assured results a longer period. The most important experiments may be consummated, and the most essential information acquired in a single year; while other results may be added, and those of the first season verified or corrected, by trials continued through a series of subsequent years.

The farmer who is accustomed to experimenting on a limited scale, with reference to but one, or a few points of inquiry, does not perhaps realize how greatly the results may be enlarged, with but little extra labor. By introducing additional elements into the investigation, and by properly combining them, the effects may be multiplied in a ratio equally surprising and profitable.

If, for example, he plants a portion of his cornfield without any manure whatever, and then adds separately to other successive portions of the same field the various fertilizers in general use, that are known to contain one or more of the elements of maize, he performs a very usual and doubtless an instructive experiment, and the greater the number and variety of fertilizers employed, the larger will be the stock of information acquired.

But this, however useful, is still a limited and partial investigation. The experiment may easily be extended, so as to render it much more comprehensive and valuable. Let us suppose the fertilizers he has selected to be ten in number. Then, by applying each of these in three different and distinct quantities, the number of effects will be materially augmented, and the knowledge acquired will be greater in amount, as well as more accurate and more valuable. He will not only discover which are the best manures to apply, but will also obtain some useful hints as to the proportion of each required.

Again, he may still further extend and vary this investigation, by applying the several fertilizers in three different modes, viz. : 1, by ploughing them into the ground before planting ; 2, by placing them in the hill or drill at the time of planting ; and 3, by combining these two methods into one. This would again multiply the whole number of results, and greatly increase the total sum of acquired knowledge. If the number of fertilizers, which is assumed to be ten, be multiplied by three, and that product by three again, it will show how many points of information would arise from such a combination of experiments.

To make this clearer, we will suppose that he appropriates to each fertilizer several rows through the field, amounting to two square rods of ground; making, when the fertilizers are all applied, twenty square rods. He next applies, on the adjoining twenty rods, the same fertilizers in larger proportions; and again, on a similar section, the same fertilizers once more in still larger quantities. He now has sixty rods planted, and thirty different conditions of manure. Thus far, however, the applications have all been made in one way only. The manures have been ploughed into the soil before planting. On the next sixty rods, therefore, he duplicates the amount already planted, making no change, except that the manures are now applied in the drill. Finally, he plants a third section of sixty rods, in the same manner as before, with the exception that the fertilizers are applied differently, by combining the two previous methods into one. He now has his corn growing under ninety different conditions of fertilization, on one hundred and eighty rods, or a fraction over one acre.

In whatever way these experiments may each one terminate, if they have been rightly performed, his object is gained. The results, it is true, may not all be equally definite and certain; this is not to be expected. Yet he derives some hint, or information, more or less plain and positive, from each separate application, while in many instances the instruction is clear and unmistakable as language can make it.

Some of the fertilizers employed will perhaps add nothing to the yield; *showing that the constituents of corn contained in them were already present in the soil in suitable amount and condition.* Others will add to the product in various proportions; some of them increasing the yield probably fifty per cent. or more as compared with the product on the unmanured ground.

A careful comparison of all the results, and of the ratio they bear to that of the unfertilized section of his field, will teach him which of all the fertilizers employed contain those precise elements of corn that were either absent from the soil, or, if present, were deficient in quantity or availability.

Before this trial was made, he did not know, and could not have predicted, the precise effect in any one instance out of ninety. He now has, if the experiments have been *carefully* and *accurately* executed, an intelligible result for each condition. With proper caution in making his deductions, he may derive from this experimental crop an amount of instruction and practical knowledge that could not have been obtained from any other source.

Even though some of the results should appear doubtful, and some of his deductions prove erroneous, there would still be a clear and decided preponderance of positive and reliable information that would pay him many times over for the extra time and labor it has cost him.

He may not have achieved a very remarkable crop, as to the aggregate number of bushels, but he has accomplished a more important object. He has not been aiming at a large present yield. He has merely been laying the foundation for many bountiful and remunerating crops during many years to come. Still the chances of a large product are all in his favor, even for the current year.

It is not only probable, but nearly certain, that, while he has been solving questions of permanent importance to his farm and to his future crops, he has at the same time obtained more than an average yield. While gathering an ample harvest of corn, he has gathered along with it a still more ample harvest of valuable information.

The trial crop here described, and the experiments embraced in it, are suggested, as one out of many plans, that will doubtless occur to the mind of the practical farmer. Those who find the subject of sufficient interest, will very likely be able to improve upon these hints. But the one essential idea that the author desires to impress upon the mind of the farming reader is, that the *system* here illustrated is capable of great expansion, and of an infinite variety in its application.

Single and isolated experiments, however useful in themselves, give no adequate idea of the increased effect that may be produced by a series of them, when ingeniously combined and accurately performed. In the hands of a skilful cultivator, a true method or system of experiments may become an invaluable instrument of knowledge and of power; for there is scarcely any kind or degree of needed information which it may not be made to develop, and few practical problems in agriculture which it will not help to solve.

PREPARATION OF THE SOIL.

In preparing the ground for corn, the subject requiring the farmer's earliest and most careful attention is disintegration. To impart to the soil, before planting, a suitable tilth and mellowness, by mechanical processes, is an indispensable preliminary. The means of doing this, and the methods practised, are various, and of different degrees of merit; but the amount of disintegration they are capable of imparting is the great and leading consideration. The instrument, or the practice that will most completely effect the pulverization of the soil, carrying the subdivision of its particles nearest to the point of ultimate possibility, is the one to be adopted by the cultivator.

In every branch of husbandry, yet in none perhaps so much as in corn culture, the thorough reduction of the earth by mechanical division and subdivision is a matter of primary and fundamental importance.

There are, it is true, exceptional cases requiring a

different treatment, but deep, thorough, and repeated ploughing is the great general rule, and the exceptions are comparatively few.

Land that is naturally sandy and porous, with a subsoil of like structure, rendering it incapable of retaining manure, requires, of course, another method. It demands, in fact, not so much a different mode of culture, as an entire change in its condition. A liberal addition of clay, ashes, and marl of the right kind, either or all in due proportions, followed with stable-manure and green crops ploughed under, would in time reconstruct such a soil, and would probably pay well for the process. But apart from such instances as this, it is perfectly safe to advise a more frequent, careful, and accurate use of the plough than that commonly practised.

If, on the other hand, the soil intended for corn is naturally wet, with a subsoil impermeable to water, it must be under-drained. This treatment is simply a matter of necessity, and cannot be superseded by any other. Even in most of the ordinary soils, it is the opinion of many farmers that under-draining pays well in the long run. But, in such a case as the one under consideration, it is not merely advantageous, it is indispensable; and to attempt to raise corn, or any other important crop without it, is a criminal waste of time and labor.

The *Working Farmer*, for May, 1861, has some useful suggestions for the treatment of the ground in corn culture. In reference to the first breaking up of the soil, the writer remarks :

"This should be performed by running the surface plough to full depth, and following with a lifting sub-soil plough, the latter propelled by a separate team, with its beam in the bottom of the furrow left by the surface plough, and not skating along the surface, merely scratching or slightly disturbing the bottom of the furrow. This lifting, subsoil plough not only under-cuts the land side so as to enable the next furrow-slice to break off more deeply and pulverize more completely, but at the same time it lifts the previously turned furrow-slice for a short distance, perfectly disintegrating its particles; for the resolution of its forces being upward and outward, renders all the soil above it, like that above the mole-track, perfectly divided."

Nearly all the large crops we have any account of, have been produced, to a large extent, by thorough tillage. Manures are doubtless highly useful,,and have their share in producing results. But it is tillage, beyond any doubt, that gives to fertilizers their greatest value and effect.

The true philosophy of thoroughly aërating the soil, so that it may not merely admit, but invite, the approach of air and water to the growing roots, is sufficiently shown in the fact, that the chemical elements of water and of air constitute ninety per cent. or more of nearly all growing plants.

In addition to this, it is to be remembered, that the plant-food already in the soil, as well as that applied by the farmer, depends upon the action of these same agents for its availability and nutritive effect. Of all the fertilizing elements contained in the earth, or added to it, there is not one that can produce its proper and legitimate result in supplying food to the growing plants without the presence and influence of either air, or water, or of both combined.

These facts are well understood, and clearly indicate the necessity of facilitating, by every possible means, the access of descending rains and of atmospheric influence to the roots of growing corn. But in order to accomplish this, the earth must be brought to a proper condition before the grain is planted. The soil must be made mellow and porous, by deep and searching processes of pulverization often repeated.

It cannot, then, be too frequently or forcibly suggested to the agriculturist that, the more he contributes to break up, crush, grind, triturate, and subdivide the particles of the soil, before planting, so much more does he coöperate with Nature, and assist her generous efforts to return him a liberal yield.

In thus dwelling, with some repetition, upon what is deemed an important subject, we may perhaps weary the patience, or provoke the severity, of some critical reader; yet such is the consequence of this principle, and such the extent of its influence, that if we could thereby impress it more effectively on the minds of our cultivators, we would not hesitate to employ yet a dozen more terms to express the same idea, did the language contain them ; for there is no reason to doubt that, if a more thorough system of tillage were practised by every one of our four million farmers, it would add to the corn-crop of this country, in a single season, many million bushels.

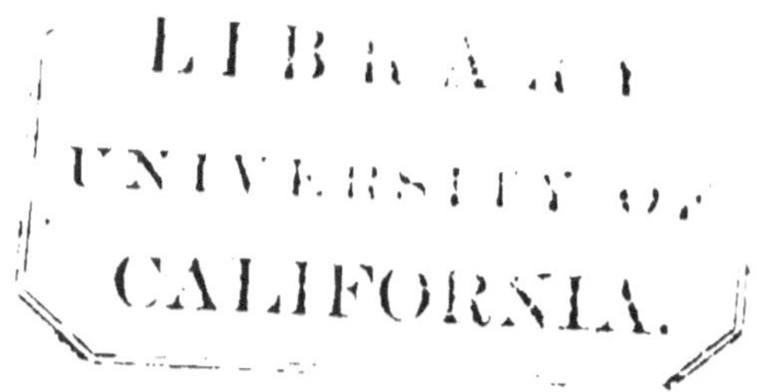

MANURES.

THERE is no grain crop in this country that so well remunerates the cultivator for a liberal application of manure as Indian corn. Although it is capable of a fair and sometimes even a generous yield on indifferent or unmanured soils, it is but short-sighted economy, on the part of the husbandman, to take advantage of this fact, by attempting to raise it without enriching the land. If the object of the agriculturist is to get the largest possible return for the manure applied to his ground, he will effect it more certainly by a generous allowance to the maize crop than in any other way.

The fertilizing materials that may be usefully applied to the cornfield are so numerous, so various, and many of them so readily procured, that no cultivator is justifiable in neglecting to apply them on a liberal scale.

The standard manure for Indian corn, as well as for other crops, is undoubtedly that of the farm-yard and the stall. Nature has ordained that domestic animals, which consume so largely the products of the earth, shall in some measure compensate the proprietor, by supplying him with the best and surest means of restoring its fertility. Yet this supply is not alone sufficient for the requirements of the soil, and the farmer finds it necessary to have recourse to other sources, which are fortunately neither few nor inaccessible.

After exhausting the contents of the cattle-yard and the compost heap, or, what is perhaps still better, in connection with these, he may employ, and often with great advantage, some of the various fertilizers in the market. In doing so, however, great caution is needed to avoid the impositions continually practised by the venders of worthless adulterations. There are several of the commercial manures composed of such articles as nearly all farmers either have or can readily and cheaply procure; and many have adopted the habit of preparing these on their own premises. There is no good reason why this practice should not be universal. The man who uses fertilizers prepared by himself is always sure of their quality, and will generally find them less expensive.

The following enumeration embraces most of the fertilizing materials in general use for the corn crop as well as some that are not usually employed, though they might be, in many cases, with advantage:

1. The manure of the FARM-YARD, comprising the excrement, solid and liquid, of horses, cattle, and other stock, and also the decomposed vegetable matter combined with them. The latter includes straw, weeds, leaf-mould, swamp-muck, and every variety of vegetable substance, which, if well managed, will

not only largely increase the aggregate amount, but will be fully equal in value to the best animal manure.

2. POUDRETTE, or the various preparations of night-soil. This is a highly concentrated and valuable fertilizer. The simplest, and perhaps the best mode of preparing it, is to combine with the night-soil a liberal proportion of dry mould, charcoal-powder, or sulphate of lime (gypsum). These may all three be added with excellent effect. Home-made poudrette, when rightly prepared, is much superior to the commercial article.

3. The various GUANOS, of which the Peruvian is by far the best. The powerful nature of this fertilizer requires caution in the use of it. In solution it is found useful for steeping, and is also applied as a liquid manure.

4. BONE-DUST.—The value of this fertilizer, for *immediate* use, depends in a great measure on its being finely ground. By the usual mode of grinding it, the effect, though more lasting, is comparatively slight the first season. The Flour of Bone is a finer preparation than the other, and though more costly, is far better for immediate effect.

5. SUPER-PHOSPHATE OF LIME, or vitriolized bones. The immediate value of bone-dust is increased, and the effect rendered much more speedy, by converting it into super-phosphate of lime. This is done by adding to the ground bones from one-half to one-third of their weight of sulphuric acid (according to the strength and purity of the acid), with a like quantity

of water. But an equal effect may be obtained, at a less cost, by decomposing ground bones with green manure or swamp-muck.

6. WOOD ASHES, leached and unleached. The former, though less valuable, still retain most of the constituents of the unleached, having lost only a portion of their soda and potash. In either form, ashes are a most useful fertilizer, and adapted to nearly every description of soil.

7. PLASTER, or sulphate of lime. Plaster is the name given to ground gypsum. It is generally beneficial to corn, and sometimes in a remarkable degree; its effect depending very much on the character of the soil.

8. LIME, oxide of calcium. That obtained from burnt shells is by many considered superior to any other. The best results from the use of lime are found in soils that abound in vegetable matter. This material is found to be much better applied in small quantities, occasionally repeated, than in large quantities at one time.

9. SALT, chloride of sodium. There is much difference of opinion in regard to this fertilizer, but there are doubtless soils on which it is useful. It has a tendency to check the growth of weeds, and its effect on grain is to increase the solidity and weight.

10. LIME AND SALT MIXTURE.—This may be prepared by adding two parts of lime to one of common salt, or by slacking the lime with a saturated solution of salt. The preparation should be made several months before using.

11. NITRATE OF POTASH, saltpetre. The effect of this fertilizer has been found in some instances quite remarkable; but like most other manures, it varies with the soil. It makes an excellent solution for steeping.

12. NITRATE OF SODA.—For soils deficient in soda, this application can hardly fail to be useful. It is sometimes applied in connection with the sulphate of soda, with an increased effect.

13. SULPHATE OF AMMONIA. }
14. PHOSPHATE   "   } All growing plants require ammonia, and what they do not obtain from the atmosphere by the agency of descending rains, must be derived from the soil, or from the manures applied to it. Hence any fertilizers containing this principle may be applied to Indian corn with un-doubted advantage.

15. PHOSPHATE OF MAGNESIA AND AMMONIA.—This compound is highly commended by Professor Johnston for its marked effect upon Indian corn. He cites a case in which three hundred pounds per acre increased the crop of grain six times and the stover three times. "It is prepared by pouring mixed solutions of sulphate of magnesia and sulphate of ammonia into a solution of the common phosphate of soda."

All of the above fertilizers contain a greater or less amount of the constituents of maize, and are therefore adapted to that crop, though in different degrees. Which of them may be used to the best ad-vantage in a given case, or how many of them, or in what proportions, are questions to be determined chiefly by the character of the soil.

If the farmer has ascertained the requirements of his soil; if he has determined, either by experimental processes, or otherwise, in what constituents of maize it is deficient, he is then prepared to apply his fertilizers with intelligence and effect, and so far as it depends upon the mere presence of enriching material in the earth, he will easily be able to bring his land up to any capacity of yield he may choose, being only limited by the expense.

He will, however, discover that the mere presence of manures is not all that is required, even though they contain the precise ingredients that are lacking in the soil. The condition in which they are applied has no small influence on the effect they are capable of producing. If they are in a hard, concrete, undivided mass, they should be pulverized. If they are not, indeed, already in a state of minute subdivision, they should be brought to that condition before applying them. Some of the saline fertilizers are procured in a state of powder, others in hard lumps that need to be finely crushed or dissolved.

But the manure requiring most attention in this respect is that of the farm-yard. It is not a little remarkable that in the very case where the process of reduction and disintegration is most of all needed, it seems to be most neglected. The contents of the stalls and of the compost heap, which, from the variety of materials they comprise, need to be elaborately worked over and subdivided, in order to be thoroughly *intermingled*, are yet frequently carted upon the land in rude lumps and unbroken masses that strangely contrast with the fine roots and fibres through whose minute mouths they have yet to enter before they can nourish the growing corn.

"Few farmers," says the editor of the *Agriculturist*, "comprehend the importance of attending to this item in the preparation of their stock of fertilizers. They are often carried to the field in the spring, in the coarsest form possible, the hay and straw not fermented at all, and the coarse clods carried in to the yard last summer, not broken. They are spread in this state, and the large lumps are ploughed under so that they are not immediately available for the sustenance of plants. Plants feed mainly at the extremities of the rootlets, through mouths too small to be seen by the naked eye. The finer the manure is made, the more easily it is dissolved in water, and the sooner it passes into the circulation."

The cultivator who intends to secure a maximum crop, or even a tolerably liberal and paying yield, will find it necessary to attend to his fertilizers, whatever may be the kinds employed, and to reduce them to a suitable degree of fineness before applying them to his soil.

But, in order to secure to the growing plant the full and legitimate effect of the manure applied, there is still another condition remaining to be complied with. The fertilizer and the soil require to be intimately *blended*. It is not enough that they are, each of them, completely and thoroughly pulverized; they must also be, and with equal thoroughness, intermingled. The particles of manure must be effectually and uniformly distributed among the particles of the soil.

Prof. Way, in a lecture before the Royal Agricultural Society of England, finely illustrated the relations of the soil to the plant that grows in it, by comparing the former to the stomach of an animal, observing that Nature had given to the soil the function or office which in animals is performed by the gastric juice and the chyle—that of preparing and digesting the food of plants. Nothing can show plainer than this analogy the importance of incorporating fertilizers with the soil.

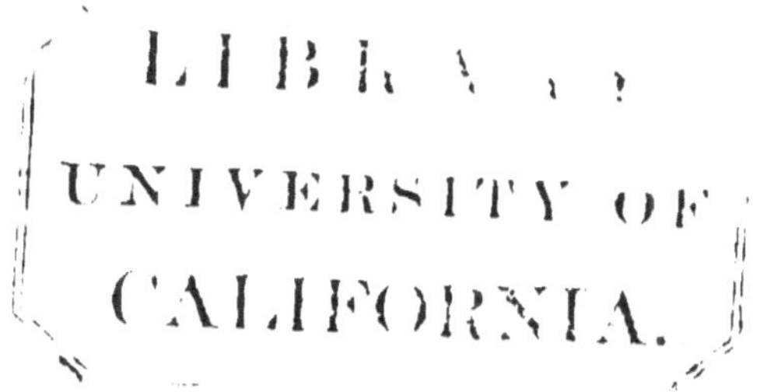

PLANTING.

HAVING selected his seed-corn with discriminating care, having prepared it by steeping for an early and vigorous start, having given it a moderate coating of tar to shield it from its earliest enemies, and finally, having imparted to his soil the requisite degree of mellowness and fertility, the farmer is now prepared to commit his seed to the earth.

But here again he is confronted by problems peculiar to the soil, and for the solution of which he must rely mainly upon his own investigations. Before depositing his seed in the earth, it is needful to determine the proper depth for planting, and the proper intervals of space. These are points that depend materially upon the variety of corn, the character of the soil, and the manner of treating it. There is, therefore, no fixed or uniform rule on the subject. The depth for planting varies from one inch to two or three. In a very heavy soil, the former would perhaps be sufficient; in a very light soil, the latter would scarcely be too deep. But the proper distance between the grains is subject to still wider latitude, and is even more dependent upon varying circumstances.

The best advice, then, that can be given to the cultivator in this case, as in a previous one, is to carry his inquiries directly to the soil, and obtain his answers there. All the information necessary for his purpose, in regard to these points, he can obtain in a single season, by a series of well-managed experiments.

There are two modes of distributing the grain in planting, in regard to which agriculturists are divided in opinion and practice, some maintaining that planting in hills is most successful, while others are equally strenuous in favor of drills. The preponderance of opinion, however, is in favor of the latter method. Our own experience is entirely in favor of drills, which seems to be the mode of planting that will secure the largest product. Still this point, like most others in husbandry, is one that every farmer can determine for himself. But let him adopt which he may of these methods, the same question of spaces remains to be solved.

This investigation is somewhat complicated, and resolves itself into two inquiries:

1st. What is the average distance between the grains, or, in other words, what is the area of soil to each grain, that will give the largest yield per acre? Now, when this is determined, it will be found that there are various modes of distribution that will give the same area to each grain, and yet no two of these would probably give the same result per acre. For instance, suppose it were ascertained that three square feet of ground to each grain would give a larger yield than any other area. Now a distribution of hills, three feet apart, with three grains in the hill, would satisfy this condition. So also an arrangement of drills three feet asunder, with stalks twelve inches apart in the drill, would equally fulfil the condition. But the product per acre of these two methods would not be the same. Here, then, arises the other inquiry:

2d. With a given area to each grain, what is the arrangement or distribution of the grains that will give the largest product per acre?

This problem deserves the attention of every agriculturist, for it determines, as elsewhere stated, the limit of possible yield. The solution of it can undoubtedly be unfolded by the method of experiments, if they are well planned and carefully executed.

The intelligent and thoughtful farmer understands that, as there are many modes of distributing the grain in planting, he cannot expect to adopt the best, without knowing which it is, and this he cannot know without making a trial. He therefore determines to vary his modes of planting, remembering that the greater the number of plans he tries the more certain he will be of finding out the best. Accordingly, he plants a part of his field in hills, and part in drills. The former he places at different distances asunder, varying at the same time the number of grains in each. The drills are in like manner placed at different distances, and the intervals between the grains are also varied. In all this there is no great intricacy and no real difficulty. He thus examines, with but little extra trouble, some fifteen or twenty different arrangements for planting; and as all these trials are introduced into his regular crop, they involve no interruption of his general plan. If, now, he should find that some one or more of these methods give him a yield of eighty or ninety bushels per acre, while the plan he is accustomed to has seldom given over sixty bushels, he would be very likely to open his eyes to the value of such experiments.

But there is one essential thing to be observed and remembered. The more closely the grains are planted the more the soil is to be enriched, and the more thoroughly and deeply must it be tilled. It is also to be observed that the small varieties admit of closer planting than the large.

To guide the cultivator in pursuing the investigation of this subject, the table given on pages 130 and 131 may be of some service. It exhibits twenty-one different arrangements for planting, with three several results for each per acre. These results are given in bushels,* omitting fractions.

The average weight of shelled corn per ear is about five ounces. In the table, therefore, three ounces are taken as the estimate for a small ear, five ounces for one of medium size, and seven ounces for a large ear. The fifth column indicates the yield, supposing the stalks to contain, on an average, but three ounces of corn each; the sixth column gives the yield for five ounces, and the last column for seven ounces.

---

* The bushel is taken at 56 lbs., that being the legal weight in most of the States.

6*

TABLE OF RESULTS

For Different Distances in Planting.

HILLS.	1	2	3	4	5	6	7
					Bushels per Acre, with one Ear to Each Stalk.		
	Distance apart.	Stalks per Hill.	Square inches to each Stalk.	Stalks per Acre.	One Small Ear.	One Medium Ear.	One Large Ear.
1........	24 in.	3	192	32,670	109	182	255
2........	24	4	144	43,560	145	243	341
3........	30	3	300	20,908	70	116	162
4........	36	3	432	14,520	48	81	114
5........	36	4	324	19,360	64	108	151
6........	42	4	441	14,223	47	79	111
7........	48	4	576	10,890	36	60	85

TABLE OF RESULTS—(*Continued.*)

DRILLS.	1	2	3	4	5	6	7
					BUSHELS PER ACRE, WITH ONE EAR TO EACH STALK.		
	Distance apart.	Stalks apart.	Square inches to each Stalk.	Stalks per Acre.	One Small Ear.	One Medium Ear.	One Large Ear.
8.........	24 in.	6 in.	144	43,560	145	243	341
9.........	24	12	288	21,780	72	121	170
10........	30	6	180	34,848	116	194	272
11........	30	9	270	23,232	77	129	181
12........	30	12	360	17,424	58	97	136
13........	36	6	216	29,040	97	162	226
14........	36	8	288	21,780	72	121	170
15........	36	9	324	19,360	64	108	151
16........	36	12	432	14,520	48	81	114
17........	40	4	160	39,204	131	218	305
18........	40	6	240	26,136	87	145	204
19........	40	8	320	19,602	65	109	153
20........	40	9	360	17,424	58	97	136
21........	48	12	576	10,890	36	60	85

The mode of spacing given in the second and eighth lines of the above table, allowing but one square foot of soil for each stalk, is introduced here for the purpose of comparison only, and not with a view of being attempted in practice.

In the first and ninth lines, also, the arrangement is too much crowded for general field culture, but may be well enough introduced in a series of trials; though the ninth method is, in fact, the same as that practised by Major Williams, of Kentucky, who succeeded in getting one hundred and sixty bushels per acre, which is only ten bushels short of the maximum result given in the table for that method.

The mode of distribution given in the seventh and twenty-first lines will probably yield the largest ears, but not as large an aggregate product as some of the others.

A few of the results in the above table are such as no practical agriculturist would expect, in the present state of our knowledge, to be able to arrive at. What may be hereafter accomplished, when the genius of our farmers shall have introduced and perfected new, and at present unknown, varieties of corn, and when science and skill shall have more fully developed the higher possibilities of the experimental system, it would be difficult now to say. But, in a soil favored by nature, and rightly improved, there is, we think, no impossibility in obtaining any of the results of the foregoing table, with the exception of those given in the first, second, and eighth lines, and, in the last column, of the tenth and seventeenth lines.

No man can tell what his own particular soil is capable of, or can be made capable of, until he has proved it. The cultivator who has always pursued one invariable method, without trying or examining any other, can never be sure that his own method is the best, or that it is anywhere near the best, or that it is even comparatively a good one. He may have been unconsciously planting his corn for years upon a wrong principle, which a few simple experiments would have long since corrected. It is quite possible that he has been losing some ten or fifteen bushels of corn per acre annually, for years, only for the want of a little more knowledge, which might have been acquired with a little more trouble.

In order to determine this point, let him submit his method to a rigorous investigation. Let him compare it with other methods, in various and repeated trials. Let him put himself in communication with Nature, and in a series of careful and patient manipulations, he will be able to draw out from the bosom of the earth a generous revelation of the laws that regulate her hidden treasures. By a system of experiments well framed and faithfully carried through, he will be able to pour a flood of light into his soil that will disclose unsuspected mines of cereal wealth.

In regard to the other details of planting, they are few and simple. Great precision is necessary in marking out the rows, to have them as regular and straight as possible, in order to facilitate the after-culture. It is the practice of some farmers, and well worthy of general adoption, to use the subsoil plough in striking out the furrows. "From the peculiarity of this plough," says Prof. Mapes, "the soil will be left in a much more divided condition than by the simple ploughings alone, besides the fact that this fresh disintegration gives strange germinating power to the soil in which the seed is now to be introduced. This lifting subsoil plough will affect the soil at the surface for one foot each side of its line of travel, so that the after-culture between the rows need not approach so nearly to the corn."

It is a good rule in planting maize to put more grain into the ground than is intended to remain. It provides against casualties, and can be thinned out at the second hoeing. It has also this advantage, that it enables the cultivator, at the time of thinning, to make a selection. There is often at that stage of growth a marked difference in the plants; and it is an important point gained when the stalks are so abundant that all the small and inferior ones can be rejected, and still leave an ample supply of large, healthy, and vigorous plants. This certainly increases the chance for a good crop, and seems entitled to more attention than it has usually received.

Another point connected with planting, and too important to be overlooked, is the uniform covering of the seed. If this is not properly attended to, there can be no uniformity of depth, nor equality of growth. In planting by hand, it is scarcely possible to accomplish this object. A variety of planting implements, of more or less merit, have been introduced within a few years, and no good farmer should be without one. In drill-planting, this implement is still more indispensable than for planting in hills, and very speedily reimburses the outlay in the saving of time and the superior accuracy of the work.

AFTER-CULTURE.

I_F the ground intended for corn has been prepared before planting, in the thorough manner indicated in a previous chapter, the labor of after-culture is thereby diminished. The more mellow and porous the condition of the soil at the time the grain is put into it, and especially if it has been deeply disintegrated by the subsoil plough, the less deeply and frequently will it require to be disturbed during the growth of the plant. A certain amount of tillage is, of course, indispensable to keep down the weeds, and to facilitate the access of air and water to the roots. But the true theory of after-culture is doubtless to keep the earth that surrounds and covers the roots of the plant as open, and loose, and porous as possible, without, at the same time, doing violence to the roots.

Hence it is obvious, that if the soil is brought completely into this condition at or before the commencement of germination, it will not require the same amount of disturbance afterwards, with the plough and other implements, that it must necessarily demand in those cases where the roots, and stems, and minute fibres are compelled from the start to struggle through a hard, compact, and neglected soil. All the tillage, however, that can be given with safety, and all that the earth really needs, in order to keep it aërated, and to prevent the growth of weeds, it must have.

There is no greater enemy to the maize-crop than weeds; and it is even doubted by some whether all the other enemies of this cereal combined accomplish so great an amount of mischief as these spontaneous and all-pervading pests of husbandry. They are sometimes kept out of the cornfield by precautionary measures, if these are early adopted; but this cannot always be effected with certainty. There are certain fertilizers that have a favorable tendency in this direction, and especially common salt, which, on some lands, produces the twofold effect of increasing the crop and checking the growth of weeds. But when, in spite of all the precautions that can be employed, these plagues and persecutors of the soil obtrude themselves into the cornfield, they must be dealt with promptly, at whatever sacrifice. They must be extirpated at once, even if it is done at the risk of some damage to the roots of the grain. The growing corn can better afford to encounter the possible loss of some portion of its roots, than to endure the presence of these greedy interlopers, that swarm into the field, only to rob the soil of its nutriment, and to exclude from it the genial sunlight.

The germs from which this infinite variety of weeds annually and spontaneously springs, are not the product of a single season, but doubtless the gradual accumulation of many years. Hence the farmer's only certain and final delivery from them is in the constant and complete extinction of them every season, before they go to seed, for a series of years, until the last lingering germ is developed and destroyed.

It is thought by some that a certain amount of root-pruning is no disadvantage to corn, but rather a benefit; and that, consequently, the plough cannot be too much or too often used in after-culture. Now it is not, perhaps, impossible that root-pruning may be, to some extent, an advantage, though the weight of opinion is clearly against it; but, supposing it to be in some degree necessary, this scarcely justifies the excessive use of the plough after germination has commenced. If root-pruning be at all, or under any circumstances, salutary, it can only be so within certain limits, and when performed with judgment and care. But the action of the plough is necessarily violent and indiscriminate, rending with fatal energy whatever resists its progress. The very qualities that give to it its greatest value, would in this case impart the greatest mischief. When rightly used, it is an instrument powerful for good. But when driven among the rows of young and tender corn, in the capacity of root-pruner, it becomes an agent of destruction.

The horse-hoe and the cultivator are less objectionable than the plough; but after the corn is well up and under good headway, with its roots ramifying the soil in every direction, even these implements are more or less perilous, and should be employed with the utmost caution. Even while the sprouting grain is scarcely yet four inches above the earth, its industrious roots have already radiated to an amazing length, and some of them have doubtless crossed the track of the plough.

It is the opinion of some agriculturists, that a hand-cultivator might be so contrived as to accomplish all that is needed in the after-culture of corn. The common garden cultivator is said to have been used in some cases with entire success, where the soil had been rendered extremely mellow before planting. If some ingenious modification of the hand-cultivator should be found adequate to all the requirements of after-culture, it would indeed be a great gain to the cornfield, provided the labor of propelling it were not so great as to form a serious objection.

It is certainly not unreasonable to suppose that an implement, combining the merits of the horse-hoe and garden cultivator, might be constructed on a scale of size fitting it for hand use, that would answer every needful purpose for after-culture, in all cases where the soil is deeply and sufficiently pulverized before planting. Nor is there any good reason why it might not be so contrived as to do the work of the hand-hoe also. Such an instrument, if successful, would be found equally applicable to root-crops, and to all other plants requiring after-culture. One great advantage of this invention in corn-husbandry would be, to permit the small and prolific varieties of corn, such as the Browne, to be planted in closer drills than the horse-hoe allows, thus enabling them to reach a higher yield per acre than they can in any other way.

One man, by this invention, would do the work of a man and horse, and do it more accurately, and with less destruction; nor need it be any more laborious, with a well-made instrument, and in well-prepared ground, than the twofold operation of driving the horse and properly managing the implement he draws.

The useless habit of piling up the earth in cone-shaped hills around the stalks of corn, which was at one time almost universal, is now generally disapproved and wisely abandoned. It was formerly supposed to aid the stalk in resisting the effect of severe gales, but experience has proved this to be a mistaken notion. There was also an imagined advantage in drawing up the earth around the roots; but here again experience has developed the sounder philosophy of allowing the roots to find the earth, as they require it, by their own spontaneous movement. This they will be sure to do; and they will find the manure also, provided both manure and soil have been sufficiently pulverized and blended.

The editor of the *Cultivator*, as quoted by Emerson, in his Encyclopædia, has given the following opinion, as to the practice of hilling corn, and also as to the use of the plough in after-culture:

"All or nearly all the accounts we have published of great products of Indian corn, agree in two particulars, viz.: in not using the plough in the after-culture, and in not earthing, or but slightly, the hills. These results go to demonstrate that the *entire* roots are essential to the vigor of the crop, and that roots, to enable them to perform their function as Nature designed, must be near the surface. If the roots are severed with the plough, in dressing the crop, the plants are deprived of a portion of their nourishment; and if they are buried deep by hilling, the plant is partially exhausted in throwing out a new set near the surface, where alone they can perform all their offices."

HARVESTING AND STORING.

The mode of harvesting the corn crop differs in different sections of the country. In most of the Northern States the general practice is to cut it near the ground, when the grain is sufficiently glazed, and before the stalks begin to wither. It is highly important to determine the right stage of maturity for cutting, and requires nice discrimination as well as experience. The weight and quality of both grain and stover depend materially on seasonable harvesting. After the corn is cut and stooked, it is usually allowed to remain in the field until sufficiently dry for husking, after which the grain is conveyed to the crib, and the stalks are either stored in the barn or stacked near the cattle yard for provender. In all cases where the stover is not stored under cover, it should be stacked with great care, to secure it from being injured by the weather.

In many of the Southern and Western States, a different and more prodigal mode of harvesting is practised. The corn is there first *topped*, by cutting the stalks above the ears, while the latter are left until fully ripened, after which they are husked on the ground, and carried to the crib. Cattle are then turned into the field to consume what remains of the stover. This practice is inexcusably wasteful, and those who adopt it can hardly expect to find their corn crop a source of much profit.

"The stalks of corn," says Mr. Allen, in the *American Farm Book*, "ought never to be cut above the ear, but always near the ground, and for this obvious reason: the sap which nourishes the grain is drawn from the earth, and passing through the stem, enters the leaf, where a change is effected, analogous to what takes place in the blood, when brought to the surface of the lungs in the animal system; but with this peculiar difference, however, that while the blood gives out carbon and absorbs oxygen, plants, under the influence of light and heat, give out oxygen and absorb carbon. This change prepares the sap for condensation and conversion into the grain. But the leaves which thus digest the food for the grain are above it, and it is while passing downward that the change of the sap into grain principally takes place. If the stalks be cut above the ear, nourishment is at an end. It may then become firm and dry, but it will not increase in quantity, while if cut near the root, it not only appropriates the sap already in the plant, but it also absorbs additional matter from the atmosphere, which contributes to its weight and perfection."

Many experiments have been made on this subject, all tending to the same result, and showing that there is a gain of from five to ten bushels per acre in the amount of grain, by cutting the corn near the ground. In a trial made by Mr. Clarke, of Northampton, Massachusetts, an acre of topped corn was found to have lost between six and eight bushels of grain by the process.

But in addition to the loss of grain from this practice, there is a further loss in the stalk, which, if cut at the right season, and cured with care, forms an excellent article of fodder. The most enlightened cultivators, whose experience in the best mode of using this provender has taught them how to appreciate it, are invariably careful in securing the whole of their stalk crop, and would no sooner leave a portion of it standing in the field than they would abandon a similar amount of any other crop they raise. They would regard every ton of stover thus relinquished as a needless sacrifice, equivalent in amount of loss to the abandonment of so much hay. No sane man would think, for a moment, of gathering his timothy or clover by this "topping" process; nor is there any sufficient reason why he should leave the half of his stalk crop to perish in the field.

The usual argument in defence of this practice, that the stalks thus relinquished are not lost but consumed by cattle turned subsequently into the field for the purpose, can have but little force with any man who has seen the experiment tried. The class of farmers who adopt this improvident course and justify it by this kind of reasoning, when advised to cut their corn-fodder before feeding, invariably reply that there is no use in it; and that cattle will never half consume them, even when they are cut; thus, unconsciously, condemning their own practice. They gather the small ends and tender portions of the stover, and tell us that even these, when cut, are but partially and reluctantly eaten, and yet imagine the same cattle that turn from these with indifference, will go into a field of unharvested butt-ends and devour them.

Neither of the positions here taken is tenable. They not only contradict each other, but they equally conflict with the facts of general experience, and with the dictates of common sense. No domestic animal will eat the large ends of corn-stover, as they stand in the field, blanched and withered by the elements, while all kinds of cattle will not only readily eat them, but thrive and fatten on them, when they have been seasonably harvested, well cured, and properly prepared for feeding by such process as every good farmer understands.

Storing.—The ordinary method of preserving corn is to deposit it before shelling in long and narrow granaries, or cribs, the sides of which, and sometimes the ends, are constructed with laths or other narrow strips, so arranged as to leave spaces of an inch or more between, for the purpose of ventilation. The corn-crib should never be made more than nine or ten feet wide. If this width is exceeded, the grain at the centre is exposed to the risk of damage by heating. In all cases where greater width is necessary or desirable, it is a wise precaution, and perhaps a sufficient protection to the grain, to ventilate the crib through the centre of the flooring.

For Measuring Corn, either shelled or in the ear, the following rule is given in the *American Farmers' Encylopædia:* " Having previously levelled the corn in the house, so that it will be of equal depth throughout, ascertain the length, breadth, and depth of the bulk; multiply these dimensions together, and their product by four; then cut off one figure from the right of this last product. This will give so many bushels and a decimal of a bushel of shelled corn. If it be required to find the quantity of ear corn, substitute eight for four, and cut off one figure as before."

ENEMIES OF CORN.

THE maize crop, in its liability to the depredations of enemies, shares the common fate of the vegetable world. It is, however, in this respect, more fortunate than most kinds of grain and fruit. Its foes, though possibly as numerous, are far less fatal than those infesting the wheat crop, and some other vegetable products.

In a general survey of the combined results of disease and hostile ravages, it must be admitted that this cereal has escaped serious calamity in a remarkable degree. From disease it is so nearly exempt as to be considered virtually untouched. It has, indeed, formidable enemies, but most of their attacks can, by seasonable measures, be either mitigated or prevented. The corn crop has, also, this further advantage—that the most serious inroads upon it occur at that stage of the growth while it is not yet too late to replant.

The most common and familiar enemies of the corn crop are crows, blackbirds, squirrels, mice, and insects; the last named being entirely the most numerous, dangerous, and difficult to guard against.

The attacks of birds, mice, and squirrels can be, in

a great measure, prevented, or rendered harmless, by steeping the grain before planting (as elsewhere described) in a pungent or distasteful solution, and still more effectually by coating it thinly over with tar.

But the insect tribes are more formidable, and not so easily repelled. In order to deal with them successfully, it is necessary to understand their habits, and to anticipate their approach with well-timed vigilance.

The following are some of the insects most frequently encountered by corn either in the field or the crib: the cut-worm, the white grub, the wire-worm, the spindle-worm, the aphis mayis, the Angoumois moth, the chinch-bug, and the weevil.

The CUT-WORM is the most dangerous enemy of the corn crop, to which, however, its ravages are not by any means confined. It is an equally well-known and destructive pest of the vegetable garden. In the daytime it remains concealed in the earth, and during the night commits its ravages, completely severing the stem of the young and tender plant, near the surface of the ground. This insect is of several species, which are the offspring of moths or millers belonging to the agrotidian group. One of the most familiar of these is the gothic dart-moth, which, in midsummer evenings, makes itself unpleasantly sociable around the lighted lamp. The cut-worm, when its full size is attained, measures from one to two inches in length, the color being of various shades of gray, with the head of a brown or orange hue.

The WHITE GRUB, though often confounded with the cut-worm, differs from it in its habits, as well as in appearance and origin. Its ravages are confined to the roots of plants, nor is it ever known, like the latter, to attack the stalk above ground. Various other plants, equally with maize, are subject to its depredations; the grasses being sometimes damaged or destroyed over entire fields, in its devouring progress. The May-beetle, sometimes called the Dor-bug, is the parent of this worm. The color of the beetle is a chestnut-brown, with the breast inclining to yellow, and in length it sometimes reaches an inch, though usually a little less. The grub, as its name implies, is a white worm, with a head approaching to brown.

The WIRE-WORM.—The havoc committed by this insect is also below the surface of the ground, and extends to the planted seed as well as to the roots that spring from it. This grub is the offspring of the Elator, or Spring-beetle. It attacks, with but little discrimination, the roots of most herbaceous plants within its reach, to some of which it is often very destructive. According to Mr. Townsend Glover, the entomologist of the Department of Agriculture, "the true wire-worm is the larva of a species of elator, or click-beetle, commonly known by the trivial name of snapping-bug, from its habit of being able to throw itself some distance in the air with a sudden click, when laid upon its back; it is said to pass five years in the larva or feeding state, and resembles the common meal-worm, the body being cylindrical, very tough, of a yellowish brown color, and furnished with a distinct head, and only six legs."

The SPINDLE-WORM takes its name from its destroying the young and tender spindle of the maize. Its ravages usually commence at an early stage of the growth of the plant, while the spindle is yet but little developed. The presence of this miscreant is indicated by the withering of the leaves, which may be taken hold of and drawn out along with the spindle. A small hole may be detected in the side of the plant, near the surface of the ground, entering into the centre of the stalk, where the worm will be found—a small, yellowish insect, with the head nearly black. The moth produced from this insect, according to Dr. Harris, differs from the other nonagrians somewhat in form, its forewings being shorter and more rounded at the tip, and the hind wings of a yellowish gray. The surest way to check the ravages of these insects is to destroy them in the caterpillar state. If permitted to turn to moths, they escape, with the certainty of propagating another brood.

The Aphis Mayis, or corn-plant louse, belongs to an exceedingly numerous tribe. The aphis, of one kind or another, is found upon almost every plant in the vegeteble kingdom, and multiplies with a rapidity truly amazing. So prolific are they, according to Reaumur, that one individual, in five generations, may become the progenitor of nearly six thousand million descendants. The corn aphis, according to Dr. Harris, is found mostly below the surface of the ground, deriving its nourishment from the roots of the plant, and the crop, in light poor soils, is liable to suffer seriously from this cause. These insects, small as they are, might, by their numbers, become truly formidable, were it not that nature has placed a check upon their increase. Other insects, the enemies of these, destroy and devour them.

The CHINCH BUG is chiefly and primarily known as the enemy of the wheat crop, which, in the Western and some of the Southern States, it invades in numbers equally formidable and fatal. The chief peril of the cornfield from this gregarious foe seems to occur when the former crop is too limited in amount to satisfy its rapacity, or has been placed by harvesting beyond its reach. It has been suggested to sow clover along with wheat and other small grains, which it is thought would have the effect of detaining the insect in the field, after the grain is harvested, long enough to save the neighboring crops from its ravages. Where no such precaution has been taken, and the wheat-field has been exhausted by these greedy and pestilent vermin, if there is no other cereal at hand but maize, and the latter is within reach, it is almost certain to be attacked and destroyed, unless promptly defended by vigorous measures.

It is in the unripe stages of wheat, oats and corn, that they are chiefly liable to the attacks of these bugs. Mr. O. M. Colver, of Cedar County, Iowa, has given an interesting account of this insect in a letter to the American Institute Farmer's Club. "While feeding on the rich juices of the wheat," he remarks, "from the time it blossoms till it matures, they increase with amazing rapidity. Often whole fields of wheat, which only show a few small spots injured, are entirely killed within two weeks. Chinch bugs breed on the ground (and when it is dry many of them are in the dust) in colonies, sometimes covering one or two square feet to the depth of half an inch or an inch with bugs in all stages of development, from the tiny red insect to the black bug and up to the perfect winged insect. They commence killing the wheat nearest their colony first, but they soon widen to feet, rods, and acres. The small white spots of dead wheat in the green field show their whereabouts.

" They take their meals in clear hot days before it gets hot in the morning and late in the afternoon. They are mostly at home in the colony in the hottest part of the day, or gathered under sheaves of wheat from the heat. Do not cut wheat before it is ripe, on account of the bugs, for they only prevent it from maturing, and cutting it will do the same. They are most voracious in their growing state. I do not think they breed in oats or corn. So far as I have observed, they always attack oats after the wheat is ripe or killed by them, from the side next to the wheat. When they go from one field to another they do not commence in spots, but sweep all as they go. I have never seen them travel forty rods, from one field to another, and do any damage."

The chinch bug bears some resemblance in size, as well as odor, to another little voracious miscreant that sometimes invades the sleeping chamber. As the former insect is provided with wings, it is fortunate for the human family that it is not at the same time, like the latter, sanguiniverous.

The ANGOUMOIS MOTH.—This insect is a destructive enemy to other grains as well as to Indian corn, and sometimes commits fearful ravages on wheat, oats, and barley. It was introduced from France into this country many years ago, and is mostly confined to the Southern States, being unable to endure the climate of the North. It is only upon the ripe grain, says Dr. Asa Fitch,* that this moth preys, attacking it in the field before harvest, and continuing to work upon it in the mow and the out-door stack, but being most destructive in the bins of granaries, flouring mills, and storehouses. The eggs are laid in clusters upon the kernels of the grain, and hatch in five to seven days. The worm bores into the kernel, where it remains feeding upon the flour, until only the hull is left, whereby it appears to the eye sound and uninjured, but on being pressed is found to be soft, and by washing, the injured kernels are separated.

The WEEVIL.—There are two species of this insect, to the attacks of which Indian corn is liable, viz.: the grain weevil and the rice weevil. Both of these, like the angoumois moth, extend their ravages to the other cereals, and attack only the ripened grain, the inside of which they consume, leaving the hull entire. "In the Northern States," says Dr. Fitch, "they are mostly confined to the storehouses in our cities. They are unknown in the interior of the country, except as they have been received in seeds

---

* See his article on Insects, in the *Annual Register of Rural Affairs* for 1853.

7*

distributed by the Patent Office, which have very frequently abounded with these weevils, often to the alarm of the persons who have received them, who have been fearful a new insect enemy was being scattered over our land hereby."

PREVENTIVES AND REMEDIES.

MANY and various have been the means resorted
to for protecting the cornfield against the innumerable hosts of its insect foes. Some of these have
proved quite successful, and others sufficiently so to
encourage further efforts in the same direction. It is
by no means impossible that continued investigations
may yet teach us how to exclude from the maize crop
the most dangerous of its enemies.

Steeping the seed corn before planting, as recommended in the case of birds, though not an absolute
protection against insects, has a salutary tendency in
two ways. It is said to repel the wire-worm which
usually attacks the germinating seed, and by quickening the growth of the plant, places it sooner beyond
danger from the attacks of other enemies.

Ploughing up sward-land in the fall is attended
with advantage, by throwing out many insects from
their hidden recesses in the soil, and exposing them
to be devoured by birds, or destroyed by the frosts of
winter.

A protection against the cut-worm, sometimes

found successful, is to sprinkle a small quantity of fine-cut tobacco on the surface of the ground, closely around the plants.

The following expedient is recommended in the *Farmers' Encyclopædia:* "A pair of old wheels are to be fitted with projections like the cogs of a spur-wheel in a mill, which must be so formed as to make holes in the earth four inches deep during the turning of the wheel. The smooth track which the wheels make on the soft ground, induces the worm, in its nocturnal wanderings, to follow on till it tumbles into the pit. It cannot climb out, and is destroyed by the hot sun."

A good practice to prevent the propagation of this insect is to make bonfires in summer evenings when the moth begins to appear. Multitudes of these will swarm into the fire and be destroyed.

For the wire-worm, the following preventive is recommended by the American Institute Farmer's Club: Take of plaster and wood ashes equal parts, saturate the same with night soil from the privy vault, haul to the field in barrels, and drop half a pint in the bottom of each hill.

An expedient practised in England and recommended here, for destroying the wire-worm, consists in burying slices of potato sufficiently near the planted grain to attract the worm from it. These slices are to be examined daily, and the larvæ thus collected to be destroyed.

It is said that sowing a crop of white mustard seed will effectually extirpate the wire-worm from the soil.   Mr. Tallant reports to the *British Farmers' Magazine*, that he has freed his fields entirely from wire-worms by this means.

The chinch bug is only to be headed off by active and vigorous measures.   The following plan is reported to the *Prairie Farmer*, by H. B. Norton, of Ogle County, Ill. : " If any Western rustics are verdant enough to suppose that chinch bugs cannot be outflanked, headed off, and conquered, they are entirely behind the times.   The thing has been effectually done during the past season, by Mr. Davis, supervisor of the town of Scott, Ogle County, Ill. This gentleman had a cornfield of a hundred acres, growing alongside of extensive fields of small grain. The bugs had finished up the latter and were preparing to attack the former, when the owner, being of an ingenious turn, hit upon a happy plan for circumventing them.   He surrounded the corn with a barrier of pine boards, set up edgewise and partly buried in the ground, to keep them in position.   Outside of this fence deep holes were dug about ten feet apart.   The upper edge of the board was kept constantly moist with a coat of coal tar, which was renewed every day.

" The bugs, according to their regular tactics, advanced to the assault in solid columns, swarming by millions and hiding the ground.   They easily ascended the board, but were unable to cross the belt of coal tar.   Sometimes they crowded upon one another, so as to bridge over the barrier, but such places were immediately covered with a new coating.   The success of the defence was complete. The invaders crept backward and forward until they tumbled into the deep holes aforesaid. These were soon filled, and the swarming myriads were shovelled out of them literally by wagon loads—at the rate of *thirty or forty bushels a day*—and buried up in other holes dug for the purpose as required. This may seem incredible to persons unacquainted with this little pest, but no one who has seen the countless myriads which cover the earth as harvest approaches will feel inclined to dispute the statement. It is an unimpeachable fact. The process was repeated, till only three or four bushels could be shovelled out of the holes, when it was abandoned. The corn was completely protected, and yielded bountifully."

Broadcast applications to the land, as a means of protection against insects in general, have been frequently tried, and various substances have been employed for the purpose, in some cases with very considerable success. But the results of all such trials are necessarily affected by a variety of circumstances. Some of the articles most employed and commended are unleached ashes, lime, soot, nitrate of soda, common salt, etc. Many farmers have found advantage, as mentioned by Mr. Colman, in his third report, by mixing salt with their stable-manure before applying the latter to the land.

For the weevil and the Angoumois moth the best, and perhaps the only reliable remedy, is, as stated by Dr. Fitch, to subject the infested grain to the heat of an oven, or of a very hot room. The grain, he says, may be heated to 190° of Fahrenheit's scale without losing its germinating power, and this is sufficient to kill all the insects contained in it.

It is an interesting fact that in all this ceaseless crusade against the destructive insect tribes, Nature is ever coöperating with man. In the order of Providence, some races of the animal creation are appointed to arrest the growth and progress of others; thus limiting the results of excessive fecundity, which, if not restrained, would soon cause the earth to be overrun and monopolized by a few prolific tribes to the exclusion of all others. Natural history everywhere abounds with curious illustrations of this marvellous law, by which the equilibrium of the animal kingdom is steadily and mysteriously preserved.

"Many of the almost unheeded insects," says Levi Bartlett, in a communication to the *Country Gentleman*, "that flit about the farmer's feet, as he traverses his acres, are truly his friends and agents in destroying other species that are so injurious to his crops. The first named should be protected, the latter should be destroyed, in all their three-fold stages, as far as possible. But without some knowledge of the science of entomology, no one can discriminate, to any great extent, between his insect friends and foes. Most of the tiger-beetles should be protected by the farmer while the May-beetle, and others of his like, should be crushed beneath the foot, even if it should 'feel a pang as great as when a giant dies.'"

But of all the agents that coöperate with the farmer in his warfare upon injurious insects, there are none that render a more important service than birds. The following striking illustration of this is given in *Anderson's Recreations :*—"A cautious observer, having found a nest of five young jays, remarked that each of. these birds, while yet very young, consumed fifteen full-sized grubs in one day, and of course would require many more of a smaller size. Say that, on an average of sizes, they consumed twenty a piece. These for the five make one hundred. Each of the parents consume, say, fifty, so that the pair and family devour two hundred every day. This, in three months, amounts to twenty thousand in one season. But, as the grub continues in that state four seasons, this single pair, with their family alone, without reckoning their descendants after the first year, would destroy eighty thousand grubs. Let us suppose that the half, namely, forty thousand, are females, and it is known that they usually lay about two hundred eggs each ; it will appear that no less than eight millions have been destroyed, or prevented from being hatched, by the labors of a single family of jays. It is by reasoning in this way that we learn to know of what importance it is to attend to the economy of nature, and to be cautious how we derange it by our short-sighted and futile operations."

How plainly, then, is it the interest of the farmer to attract to his fields, to encourage and protect the feathered tribes, of every name and kind, and to wage uncompromising war against all who persecute them ; for, incredible as it may seem, there are those who find a mysterious, if not malignant pleasure in slaying these merry and innocent types of beauty. If there is, on earth, one miscreant that deserves scourging more than another, it is the shameless scamp who is so often seen prowling through fields and woods, with loaded gun, intent on the destruction of these harmless and useful friends of man, that aid in protecting his cereal treasures, while they embellish with their presence his groves and orchards, and fill the air with the music of their artless notes.

DISEASES OF CORN.

In the history of this plant disease is scarcely known. Occasionally some morbid indication, as a rust on the leaves or stalk, or an unnatural secretion, is witnessed; arising probably from wounds in cultivation, or from long-continued extremes of weather; but otherwise its history is marked with health and vigor, and it still remains untouched with any serious malady. The contrast in this respect with wheat and most other grains is so strikingly in favor of corn, as to justify the conclusion that the exemption of the latter is purposely ordered by a beneficent Providence.

The principal disease of this cereal appears in the form of a dark spongy growth, sometimes of a blue black or purple tinge, that occasionally shows itself on the stalk or leaves, but is more apt to take the place of the blighted ear. This substance increases gradually in size, sometimes reaching six or seven inches in diameter, and is generally regarded as a rank and luxuriant species of fungus. The kind of parasitic growth to which this fungus belongs, it has been found, may be in most cases effectually destroyed by an application of common salt. It has, therefore, been inferred by some that soaking the seed-corn before planting in a solution of salt, or spreading salt freely upon the surface of the ground, will have the effect of preventing this disease.

The usual theory in regard to this fungus attributes it to the bruises and lacerations inflicted upon the young plant by a reckless mode of cultivation. The bleeding that occurs from these wounds results in the formation of the dark morbid substance above described. When this happens to be in contact with the ear, it is liable to prove destructive unless discovered in season and promptly removed. When it occurs on other portions of the plant it is more or less injurious, sometimes interfering with the perfection of the grain. The only effectual remedy is speedy removal, and repeating this process as often as the fungus may reappear, which it is apt to do, and sometimes to a troublesome extent.

But this course, even when faithfully pursued, does not always insure a restoration of the plant. It is consequently a matter of importance to use precaution in avoiding the causes of this malady, and to guard carefully against the wounds and bruises liable to occur in after culture. Though some of the plants thus carelessly mangled may outlive the infliction and seem to thrive, even when the morbid growth is suffered to remain, yet a part of them must necessarily become too much enfeebled to be capable of perfecting the ear.

In addition to the above-described fungus there is

also a reddish-brown species of *rust* that sometimes shows itself on the leaves of maize. This, however, has seldom been known to extend so far as materially to affect the grain. In some instances, but more rarely, this rust has been known to fix itself on the stalk, and is then liable to produce more serious injury, and if it extends to the ear can hardly fail to diminish the product of grain.

This disease is attributed by Mr. Loraine, and some others, to the same cause that is supposed to produce the fungus, namely, the bruises inflicted by an inconsiderate cultivation. Others ascribe both this and the fungus to an atmospheric influence, or some peculiarity of the season.

But from whatever cause these maladies may proceed, the effect seems to be, on the whole, comparatively trifling, and the injury resulting has thus far proved too limited in amount to create any considerable apprehension.

This comparative and almost total exception from disease is one cause of the greater certainty of the corn crop, and is so far an argument of some weight in favor of cultivating it more extensively and more thoroughly. For the greater the degree of certainty in any crop the farmer raises, the less risk he incurs in staking upon it a more elaborate and expensive mode of culture.

THE STALK CROP.

THE stover of Indian corn, slighted as it too often is, has come to be a large and valuable item in American husbandry. Its nutritive value for feeding purposes, and the amount yielded per acre, render it intrinsically and practically an important crop. It is cultivated on three different plans.

1. It is grown primarily and most extensively as an integral part of the regular corn crop, in which case the grain is the chief object in view, the stalks holding a subordinate place.

2. It is also raised as an exclusive fodder crop, which is cured and harvested in the fall for winter use. In this case the grain, being no part of the object, is excluded by close planting, which gives a more abundant yield of the stalk. Again :

3. It is extensively grown as a green crop for cattle during summer and autumn. This process of soiling, as it is technically called, is found to be very profitable, and is getting to be largely practised.

When to these various forms of the stalk crop is added the immense supply of sweet corn extensively cultivated by the farmer for table use, we have still another addition to the aggregate yield of stalks, as well as a further contribution of grain to the general stock; thus exemplifying the manifold utilities of this cereal, which, through so many and various channels, pours annually into the storehouse of the husbandman its munificent supplies of food for man and animals.

FEEDING VALUE.—The intrinsic worth of the corn-stalk to the farmer for feeding purposes, and its nutritious quality as compared with straw, hay, and other forage, may be determined by a comparative view of the constituents of each, and also more reliably by a series of trials or experiments in feeding. As far as these trials have yet been made by practical men, the results are nearly uniform, and clearly prove the superiority of this provender. The experience of enlightened cultivators places the corn-stalk far above the straw of the other cereals in nutritive value, and justly ranks it, when properly cured and rightly treated in feeding, as quite equal to most kinds of hay. The testimony of competent judges on this subject is sufficiently clear; and the reason why any farmers are still doubtful in regard to it, the chief reason in fact why the cornstalk is not more generally prized at its true worth, is because its value is too often judged by the results of injudicious feeding, or by the unsound condition of it, arising from want of care in harvesting.

"Indian corn-stalks," says Professor Norton, "when cut seasonably and well cured, make a most admirable fodder. They are then sweet and nutritious in an eminent degree; when cut fine, and mixed with Indian meal, are eaten by cattle with much avidity, and eaten clean, butts and all. Some farmers think that really good stalks are worth about as much as the best hay. When we consider the weight of them to be obtained from an acre of heavy corn, they are probably more than equal, taking into account the respective quantities per acre."

But let us now examine the acreable product of this stover. We may then be able by a comparative view of the quality and amount of it, to form a rational estimate of its total value, and also of the proportion it bears to the value of the grain.

There are few agriculturists who know, with any degree of accuracy, how many bushels of grain, and fewer still, perhaps, who are definitely aware how many tons of stalks their maize crop yields per acre. Yet without this knowledge they can form no adequate judgment of what the crop is worth in the aggregate, and can have but a vague idea of what a bushel of the grain, or a ton of the stalks has cost them.

RATIO OF THE STALK TO THE GRAIN.—The acreable yield of the stover, and the ratio it bears to the grain, have been variously estimated by practical men. These estimates differ according to climate, variety of corn, etc. With the smaller and prolific varieties, when the yield is large, the ratio sometimes falls as low as sixty pounds of stalks to a bushel of grain. On the other hand, some of the large varieties have been known to produce, especially in warm latitudes, a growth of stalks equal to three or four times the weight of the grain. Somewhere between these extremes the average ratio is to be found, and may be computed near enough for necessary purposes.

In some estimates reported to the Patent Office from different sections of the country, the ratio was found, on an average, to be about eighty pounds of stalks to a bushel of corn. Some farmers, whose opinions are based upon careful investigation, have found the product of stalks to range from eighty to one hundred pounds to the bushel of corn. In some investigations made by the writer, the diversity was still greater, but giving a mean ratio of nearly ninety pounds of stover to a bushel of grain. But this proportion will scarcely hold good for the usual practice of cutting and curing. If, then, we assume the ratio of the grain to the stover to be twenty-five bushels to the ton, and for the more prolific varieties thirty bushels, the estimate will be found very near the average experience of farmers.

In comparing the relative acreable *values* of the grain and stalks, the case is reversed, and the grain is entirely ahead. The estimates of different farmers, in regard to the money value of the stalks, as compared with that of the grain, vary as widely as their modes of treatment. Some of them compute the stover at less than one-fifth the value of the grain, and others place it as high as one-third. When the stalks are in *good condition*, the latter estimate is probably much nearer the truth.

It will always be found that the most successful cultivators place the highest value on their corn-stalks, and for this good reason, that their method of cutting, curing, and feeding is such as to impart to them a value that many farmers have little conception of.

A few examples will be sufficient to show how the estimates of practical men vary on this subject. A farmer in Shelburn, Mass., who realized seventy dollars for his corn, computed his stalks at twelve dollars and fifty cents. A farmer of Northfield, Mass., whose acre of corn yielded fifty dollars, estimated the stalks at ten dollars. Another in Northfield, with a corn crop worth forty dollars, rated the stalks at ten dollars. Daniel Johnston, an experienced farmer of Johnsontown, N. Y., considers his corn-stalks worth ten dollars a ton; and in a crop valued at one hundred dollars, he estimated the stover at thirty dollars, and the grain at seventy dollars. Matthew Waldron, a stock-farmer of Diamond Valley, N. Y., regards his corn-stalks of more value than his best hay for cattle of all kinds, and especially for cows, and rates them at more than thirty per cent. of the value of the grain.

It has been said, and appears by no means impossible, that the stalks of Indian corn, taken in the most *perfect condition*, and converted into milk and butter, according to the best principles of feeding, may be made to realize a value quite equal to one-half that of the grain.

The quantity of stalks produced on an acre may be calculated, *when the amount of grain is known*, according to the ratio above laid down, by taking twenty-five bushels of grain to represent a ton of the stover. The quantity that an acre *is capable of pro-*

ducing may be theoretically determined in another way.

It will be found that with the larger varieties of corn, the stalks of a crop well attended will weigh, on an average, eight ounces each. But, allowing for unseasonable cutting and for defective curing, if we estimate the weight at seven ounces, it will afford a fair criterion of what an acre ought to produce. In the following table the acreable product of stover is given, according to this basis, for several different distances in planting:

	Distance apart.	Stalks per hill.	Stalks per acre.	Tons per acre, at 7 oz. per stalk.
Hills......	3½ ft.	4	14.223	3.11
	3 ft.	4	19.360	4.23
		Stalks apart.		
Drills	3½ ft.	6	26.136	5.71
	3 ft.	8	21.780	4.76
	3 ft.	6	29.040	6.35

In the above table a part of the results are perhaps larger than some farmers are accustomed to realize; but products equal to these have been obtained, and may be again.

If we now add to the above table the yield of grain, assuming it to be five ounces per stalk, the number of bushels per acre corresponding in each case to the product of stalks would be as follows:

Stalks per acre.	Grain per acre.
3.11 tons.	79 bushels.
4.23 "	108 "
5.71 "	145 "
4.76 "	121 "
6.35 "	162 "

Here it will be seen that the proportion of grain to stover is just about the same as the ratio given above, which was twenty-five bushels to a ton; showing that one estimate corroborates the other.

Let us now take a further comparative view of the stalks and grain of this cereal. Let us compare the total results per acre, including quantity and price. We will suppose the grain to bring seventy cents per bushel, and the stalks six dollars per ton. These figures are much below what the farmer should realize on a yearly average, but they will answer the purpose of illustration. We will take four different yields, viz., twenty-five bushels per acre, fifty bushels, seventy-five bushels, and one hundred bushels; assuming the stalks to be, as before given, in the ratio of one ton to twenty-five bushels. We shall then have the following result :

YIELD PER ACRE.		VALUE OF EACH.		TOTAL VALUE.
Grain, bushels.	Stalks, tons.	Grain at 70c.	Stalks at $6.	
25	1	$17 50	$6 00	$23 50
50	2	35 00	12 00	47 00
75	3	52 50	18 00	70 50
100	4	70 00	24 00	94 00

Many farmers, by converting their stalks and grain into beef, butter, and pork, succeed in realizing, by good management, a dollar a bushel for their grain, and ten dollars per ton for the stover. On an acre yielding seventy-five bushels, that would give—

75 bushels of grain, at $1 .... $75
3 tons of stalks, at $10 30
                                                  ———
Total product $105

By referring to the prices assumed in the table, it will be seen that the value of the stalks is a little over one-third of the value of the grain, and very nearly one-fourth of the total value of the crop. According to the other prices given above, the *relative* value of the stalks would be still greater; amounting to much more than one-third of the value of the grain, and more than one-fourth of the total value. In either case it is sufficiently evident how much the farmer loses who neglects his corn-stalks, and how much is gained by the prudent, intelligent man who turns them to the best account.

Cured Fodder.—When Indian corn is planted exclusively for the stover, the sweet varieties are generally preferred. It is planted much closer in the drills than the ordinary practice, and the amount of forage yielded per acre is, of course, much greater. The nutritive value is also said to be superior in this case, and the time and labor required in the cultivation are less than when the crop is raised for its grain.

In a soil naturally good and properly treated, from eight to twelve tons of cured fodder can be raised on an acre. The advantages of such a crop are therefore sufficiently apparent, as the yield is three or four times greater than that of hay, while the quality, if the stalks are well cured, is in no respect inferior. To the stock farmer this crop especially commends itself; for if his object is to winter his cattle with economy and advantage, there is no provender he can raise that is superior, for this purpose, to well-cured corn-forage.

GREEN FODDER.—The practice of sowing corn, either broadcast or in drills, for the purpose of feeding it in the green state during summer and autumn, has been gaining ground for a number of years. The advantage of this is found to be so decided that farmers are beginning to adopt it very generally. There is, perhaps, no other way in which an equal amount of nutritious feed can be extracted from the same extent of ground. All kinds of cattle and young stock thrive upon it, and for milch cows especially it is allowed by practical men to be better adapted than any other product of the farm.

This crop requires for its best results a high condition of soil, and well remunerates the application of manure and labor. Sorghum is sometimes planted as a soiling crop, but the sweet varieties of Indian corn are generally preferred. The best method of planting is in drills, which is found to give a larger yield than sowing broadcast. The product of the green fodder crop is usually from fifteen to twenty tons. Thirty tons have been raised, and higher yields are reported. Considering the amount of this provender that can be grown upon an acre, and its unrivalled excellence as a succulent food, it is not surprising that the attention of agriculturists is very generally drawn to the subject.

Mr. Josiah Quincy, Jr., of Mass., long and successfully engaged in this system of farming, estimates it is said that an acre will, by this method, and with this fodder, support from three to four cows. Mr. D. S. Curtis, of Madison, Wis., has communicated a valuable paper on this subject to the Patent Office Report for 1859, and finds also a like advantage and economy in this practice, even in a section of country where land is cheap and labor is dear.

The American Institute Farmer's Club have recorded their opinion that "nothing ever planted or sown for green or winter fodder, will give as much per acre as Indian corn;" and in the further discussion of this subject by the club, Mr. Carpenter added that one of his neighbors last year kept twenty-four head of cattle from the middle of July till frost upon two and a half acres of sowed corn, without exhausting the whole product. He believes that fifteen cows could be well kept upon one acre of corn, by commencing to cut it up as soon as it was large enough, or whenever the pasture failed, so as to keep them in a full flow of milk all the autumn.

The soiling system, when properly conducted, embraces other grains, grasses, and root crops, as well as Indian corn; but none of them contribute so largely to its success and profit as the latter; and for the simple reason that they are none of them capable of yielding so large a return in proportion to the land and labor employed.

COST OF PRODUCING CORN-FORAGE.—The cost of raising this forage has been variously estimated, but is in nearly all cases remarkably low, in consequence of the large amount per acre compared with the labor required to produce it.

Mr. S. W. Hall, of Elmira, N. Y., who obtained thirty tons from an acre, estimated the entire expense of the acre at thirty dollars, which brings the cost of a single ton down to one dollar. Mr. J. G. Webb, of Oneida County, found the expense per acre to be a little over eleven dollars for a yield of twenty-five tons, which makes the cost per ton less than half a dollar. The latter estimate, if entirely accurate, is doubtless an exceptional case.

The average cost of this crop throughout the country will probably range from one and a half to two dollars per ton; and of the *cured* fodder from two to four dollars per ton. At these figures, any farmer, who understands how to turn his stalks to a good account, must find them exceedingly profitable.

ESTIMATED VALUE OF THE STALK CROP.—In view of the increasing extent and acknowledged importance of this crop, it would be interesting and instructive to know the annual amount and value of it for the whole country. Although the census returns do not enlighten us on this point, we can still form a proximate estimate from other data that will perhaps be sufficiently accurate for a general view of the subject.

The total product of Indian corn in 1860 was

eight hundred and thirty-eight millions, seven hundred and ninety-two thousand, seven hundred and forty bushels. Allowing twenty-five bushels of grain to a ton of stover, this would give thirty-three millions, five hundred and fifty-one thousand, seven hundred and nine tons as the stalk product of the main corn crop of the country. This, however, is but one branch, though by far the largest, of the entire stalk crop. To arrive at the sum total, we must add to this the amount of stalks produced by several other crops, viz.: the crop of sweet corn, of green fodder, and of cured fodder.

Though we have not the same data to base a computation upon in the case of these crops as in the one above, we can still deduce from other grounds proximate results which, if somewhat less certain, may yet be placed low enough to give them a high degree of probability. There are four million farmers in the United States, some of whom, doubtless, cultivate all three of these crops, while others raise one or two of them, and others again perhaps not any. Now it is reasonable to presume that on an average one or another of these crops is raised by every farmer; but to bring the estimate nearer to a certainty, let us assume that by one-half the farmers in the country one or another of these crops is annually cultivated. Let us then further suppose the extent of each crop to be one acre, and the average yield of stalks four tons per acre. We then have eight million tons of stover to be added to the amount obtained above.

If now, for the sake of still higher probability, we

make a further abatement, reducing the eight million tons to six and a half million, we shall still have a grand total of FORTY MILLION TONS as the product of the stover of Indian corn for the whole country, at the period of the last census. This stover is worth, in some sections, from three to five dollars per ton; in other localities ten dollars and upwards. If we estimate the whole crop at five dollars per ton, it will give for the aggregate value of the stalk crop of the United States, TWO HUNDRED MILLION DOLLARS.

The hay crop for 1860 was about nineteen million tons. The census tables record some products of but little over a million dollars in annual value. Yet here is a product worth more than two hundred millions, which the Government has never recognized, and which, it is believed, is nowhere represented in any authentic record, or any published account.

THE ADVANTAGE OF CUTTING CORN-STALKS.—There is perhaps no question in agriculture that has given rise to more discussion than this. The opinions of farmers on the subject are as various as their practices. Some of them are accustomed to cut their stover half an inch long, others a full inch in length, while some contend that the greatest advantage is found in chaffing them very finely, not over one-fourth of an inch, and still another class maintain that there is little or no benefit in cutting them at all. This difference of opinion has kept up, for years, a lively discussion in the agricultural journals, without apparently settling any one point to the satisfaction of the opposing parties.

8*

In a theoretieal view it would appear that those who advocate the shortest cutting have taken the true rational ground. It seems to be the plain dictate of reason, and the obvious suggestion of common sense, that any mechanical process by which the food of domestic animals is *effectually* subdivided and pulverized before entering the stomach, must have a tendency to render digestion more easy, more certain, and more thorough, and that so far as this is accomplished the nutritive effect of the food is in the same degree increased.

The very teeth that nature has planted in the mouth of every animal, stationed as they are, like so many sentinels, in the entrance-porch of the stomach to guard against the intrusion of unprepared food, by arresting and crushing all that passes in, clearly indicate that pulverization is an indispensable process, and a necessary prelude to digestion.

Now, if it could be shown that the teeth alone, and unassisted, are always competent to this end, that they invariably and perfectly perform their office, never failing to reduce and grind thoroughly and rapidly every kind of food presented, there might then be some reason for doubting the necessity of cutting corn-fodder or any other provender before feeding. But it is a well-ascertained fact that the teeth are not perfect and infallible in their action—that the food of cattle, as well as the food of mankind, is very often imperfectly and insufficiently masticated. And for this there are several reasons.

It frequently happens that the impatient appetite

refuses to wait upon the slow process of mastication. The blind instinct of a voracious stomach pays but little respect to the mandibles, and the food is snatched from them before their work of grinding is fairly begun. We all know that a hungry man will often bolt his food nearly whole, and why should we expect the ox to be more of a philosopher than his master?

But it is not only in cases of extraordinary appetite that the teeth of domestic animals fail to perform their office perfectly. In all cases where the provender is by its nature hard and tough, or has a hard exterior, the process of mastication, though not entirely prevented, is seriously impeded; and whatever renders mastication slow, laborious, and difficult, must necessarily render it more or less imperfect in its results. The consequence is that the animal either abandons with weary jaws its unfinished meal, or, if the whole is swallowed, loses some portion of the benefit by the imperfect digestion necessarily resulting from insufficient mastication.

It is evident, then, that the teeth, in performing their intended function, have two serious obstacles to contend with—the hardness of the fodder and the hunger of the animal. Their efficiency is thus diminished, their work is imperfectly performed, and some portion of the food enters the stomach in a condition unfitted for its intended purpose. To meet this difficulty human skill has supplied the *cutting machine*, which, when rightly used, coöperates with the teeth of the animal in bringing the provender to the precise condition required by the organs of the stomach; rendering the mastication complete, the digestion perfect, and the animal thrifty.

These remarks and the principle involved are not limited in their application to the stover of corn, but extend equally to the cutting of hay and straw, to the slicing or pulping of roots, and to the grinding of all grains intended for domestic animals. In every case, whatever the kind of forage employed, the condition essential to the highest success in feeding is the mechanical reduction of the food to such a degree of fineness as shall render mastication easy, rapid, thorough, and certain.

This theory is not only founded in the nature of things, but is confirmed by the experience of a majority of practical farmers, as well as by the researches of science. It has been ascertained by chemists that the cellulose or fibre contained in most kinds of forage partakes of the nature of starch, being nearly identical with it, and that, when rendered soluble, it is quite as nutritious. It has also been found that the more finely this fibre is chaffed, the more soluble it becomes; and that this solubility is still further increased by the application of steam or scalding water. According to Dr. Cameron this woody fibre may be rendered to a great extent capable of assimilation, and when well assimilated or digested four-tenths of its weight may be converted into fat.

It is the opinion of many practical men, both in this country and in England, that most kinds of provender, when finely chaffed, are increased in value from forty to fifty per cent., and some consider the gain equal to much more than this. "It has been proved," says the *Working Farmer*, "that nineteen pounds of hay, cut one inch long, will take the place of twenty-five pounds of uncut hay; and it is equally true that if the hay be cut one quarter of an inch or less in length, the same relative proportion will answer the purpose. It is claimed by some that thirteen pounds of chaffed hay is equal to nineteen pounds one inch long, or twenty-five pounds in the natural state. All these facts are equally applicable to corn-stalks."

They are, indeed, even more applicable to stalks than either to hay or straw, as is evident from the nature of the case. Every consideration in favor of cutting hay becomes a much stronger argument when applied to corn-fodder.

Without insisting on all the gain in value claimed by the journal above quoted, if we assume even the half of that increase, taking thirteen pounds of hay or stalks finely chaffed as equal to nineteen pounds uncut, this will still show a gain of forty-six per cent., which corresponds with the experience of many farmers, and is rendered entirely probable by the researches of chemistry.

But in order to adopt an estimate that may be generally and certainly realized, let us put the ratio lower still. It is certainly a reasonable presumption, that on a general average the increase in the value of this stover, by chaffing it finely, will be found not less than forty per cent.; and that when it is steamed or thoroughly soaked after cutting, the whole gain will be equivalent to sixty per cent. over the value of the uncut fodder.

This principle may be more fully expressed by the following tabular form, indicating the increase in the feeding value of stalks for different degrees of treatment:

Cut to.	Gain per cent.	Gain when steamed.
1 inch.	20	40 per cent.
¾ "	30	50 "
¼ "	40	60 "

The half-inch cutting is sometimes found objectionable on account of the hard coating of the stalk being liable to get between the teeth of the animal, producing discomfort and occasionally soreness of the mouth. This, it is said, may be to some extent prevented by steaming, which softens the hard exterior of the stalk. But far the best and surest way to obviate the difficulty is to adopt the better practice of chaffing finely and steaming. It is here that the greatest advantage is found, and the greatest certainty of profitable results.

Yet notwithstanding the facts and arguments repeatedly adduced in favor of cutting corn-fodder, there are some practical farmers who reject this whole doctrine, appealing to their individual experience as a refutation of it. They assure us that cattle will sometimes leave a portion of their stalks unconsumed, even when the latter have been cut before feeding; and on this single, unimportant, misunderstood fact, the whole of their objection seems to be suspended.

It is not to be denied that some portion of the cut stover is sometimes left uneaten in the feeding-box, though such cases will generally be found to arise either from imperfect cutting, or from the unsound condition of the stalk. If the stover is not cut sufficiently small to accomplish the intended object, or if it is mouldy from imperfect curing, these causes will naturally make some difference in the amount consumed. But waiving this explanation entirely, and admitting the fact as broadly as it is asserted, that some part of the stalks will be rejected whether cut or uncut, it will yet be found that this fact itself, when rightly considered, amounts to nothing whatever as an argument against cutting or chaffing this fodder.

For the purpose of illustrating this point, let us suppose a case, and let us put the facts as strongly as possible in favor of the objector. We will suppose that some farmer, in order to test this principle, divides his herd into three equal classes, feeding one-third with whole stalks, another with the same fodder cut to half an inch, and the remaining third with the stover finely chaffed and steamed. In order to be more accurate, he weighs the fodder, giving to each animal thirteen pounds at each feeding. After every meal the feed-boxes are examined, and the quantity left is carefully weighed. At the end of a week he finds that the cattle of the first division, that were fed with the whole stalks, have left, on an average, three pounds each, out of every thirteen pounds received. He also finds that in both the other divisions each animal has left the same proportion (three pounds in thirteen) unconsumed.

From these results he infers, without a moment's reflection, that there is no advantage in cutting his stalks.

Now let us see whether this is a fair inference. His cattle have each received thirteen pounds at a feed, out of which they have eaten ten pounds, rejecting three. The primary question here is this : *What is the amount of benefit derived by each animal from the ten pounds eaten ?* That which he has not eaten, whether it were three pounds or thirty, has nothing to do with this comparison. The rejected food is a secondary matter, which, when separately considered and correctly explained, will be found to sustain rather than invalidate the theory here advocated.

According to the general principle above elucidated, that cutting or crushing the food of animals to a greater degree of fineness increases the nutritive value, it will be seen that, in the case above stated, the cattle in the second division derived more benefit from each ten pounds consumed than those in the first division ; while those in the third class, which had their stover finely chaffed and steamed, received much greater benefit from it than any of the others. In other words, the ten pounds of stover cut to half an inch were equal to thirteen pounds of the uncut, and the ten pounds chaffed and steamed were equivalent to sixteen pounds of the whole stalks.

While, therefore, the proprietor supposed he was feeding his cattle equally all around, he was virtually, and in effect, giving to one class thirteen pounds, to another about seventeen pounds, and to another nearly twenty-one pounds.

If, now, instead of fixing his mind so exclusively on the remnants, he had paid more attention to the food they ate, and to the effect produced by it; if he had pushed his experiment further, and continued the feeding a few weeks longer, weighing his cattle at regular intervals, to determine the increase of flesh resulting from each mode of feeding, he would then, indeed, have fairly tested, on philosophical principles, the theory which he now supposes he has demolished with a few pounds of remnants.

In regard to the amount of stover left unconsumed, the motive for rejection was not the same in each case. The cattle fed with the whole stalks, abandoned the last portion of them through fatigue and impatience. They found the labor of mastication too great, and the process too slow and tedious, and they gave it up in despair. On the other hand, those that received their fodder in better condition, relinquished the last part of it from mere satiety. They found their food so nutritious and satisfying, that less than the whole was sufficient for their requirements. All that is necessary to prevent waste in such cases, is to diminish the amount of food.

An animal receiving twenty-one pounds at a meal, is much more likely to leave a portion of it, than one receiving only thirteen pounds; and if a part is left in both cases, these remainders, though both significant, have each a different import. The former indicates an appeased appetite and a contented animal. The latter proclaims the incompetency of the teeth, and the animal still hungry. The former teaches the proprietor that when the fodder is rightly prepared, a less amount is sufficient. The latter gives him to understand that when the stover is fed without cutting, however small the quantity, a part will be wasted; and however large the amount, the animal will leave it unsatisfied.

There is, on the whole, but one real objection to the practice of cutting this provender, and that is the expense connected with it. What the exact cost amounts to does not appear to have been as yet very accurately determined. But without knowing this precisely, it is easy to perceive that, in any event, the expense of chaffing and steaming or soaking, is far outweighed by the advantage gained.

It has been estimated that, under favorable circumstances, the cost of cutting and soaking will not vary greatly from seven or eight per cent. on the value of the stalks. When it is considered that the value is increased by this treatment not less than sixty per cent., and in the opinion of some farmers nearly one hundred per cent., it becomes evident that the objection has no practical force, and scarcely needs to be further considered.

If the practice of chaffing and steaming the stover of corn and other kinds of forage were universal among our farmers, the effect would be to render the forty million tons of corn-stalks now raised in this country equivalent in value to more than sixty million tons; the hay crop, which is now about twenty million tons, would be virtually increased to over thirty millions, and all other fodder capable of the same treatment would be augmented in the same proportion. This consideration is perhaps a sufficient apology for the space here devoted to an examination of the subject.

NUTRITIVE VALUE OF THE COB.—The ears of Indian corn are frequently ground entire, before shelling, and the meal yielded by this process, being the joint product of the grain and cob, is found variously useful. Many farmers employ it quite extensively in feeding, and are well satisfied with the results, although there is some difference of opinion in regard to the utility of it.

As corn-meal is considered a concentrated and somewhat stimulating food, and is therefore nearly always blended with some other provender when fed to stock, there seems to be no reason why the cob may not prove to be, when ground with the grain, at least a convenient and useful divisor for reducing the pure meal.

But if the cob, while entirely free from all hurtful elements, is found to contain, at the same time, a proportion of nutritive value, then its adaptation to this object is clear and undoubted, and it becomes not only negatively useful as a divisor, but positively profitable as an addition to the feed. As chemistry has not detected any injurious quality in the cob, but has shown that it contains a positive and available nutritive value, and as the testimony of experience is mainly in its favor, there seems to be no good reason why it should not be turned to a useful account.

The ratio of the cob to the grain, when compared by weight, is found to be, on a general average, about as one to four, and the proportion it bears to the entire ear before shelling as one to five.

According to chemical analysis there are in two hundred pounds of cobs about one hundred and twenty-seven pounds of fibre, and the balance consists of various substances *capable of assimilation*, including sugar and extract, dextrine or gum, glutinous matter, a proportion of soluble fibre, etc.   There are therefore in every two hundred pounds of cobs not less than seventy-three pounds of available matter, more or less nutritive, which go to support respiration, to sustain animal heat, and are capable of being transformed into nerve, muscle, and bone.

According to this view, there are in one thousand pounds of corn and cob meal the following constituents :

Ground grain	800 lbs.
Assimilable portion of the cob	73
Nutritive matter	873
Fibre of cob	127
	1,000 lbs.

From this comparison it appears that, by grinding the cob with the corn, there is a gain of twenty-five per cent. in the quantity of food, while the nutritive matter is increased between nine and ten per cent. At the same time the general quality of the product is in some respects improved, as the new compound contains *more variety* with *less concentration* than the corn-meal alone. It is, indeed, thought by some that the addition of the ground cob to the pure meal, by rendering the latter less compact in the stomach, and therefore more digestible, contributes a value nearly in proportion to its quantity, and that consequently the corn and cob meal is worth nearly as much in feeding, pound for pound, as the corn-meal alone.

In some instances this estimate would perhaps be found not far from the truth; but it would certainly not hold good in those cases where the meal is fed for fattening purposes.

Mr. Henry A. Morgan, a New Jersey farmer, has communicated to the *Working Farmer* his estimate of the value of this feed, as derived from experiments. "I have lately," he writes, "adopted the practice of feeding corn and cob meal to my stock, and have found a very considerable advantage in it. The chemical investigation of Dr. Salsbury and the trials made by Mr. Ellsworth and other practical men, so clearly indicate the value of corn cobs, that I have been induced to make some accurate experiments on the subject. The result of these has satisfied me that corn cobs, when rightly used in feeding, are worth more than one-fourth of the same weight of hay, and nearly one-eighth of their weight of corn. As this is a matter of some interest to stock farmers, many of whom are readers of your journal, I have thought it might be well to call their attention to the subject, and it may perhaps lead to further investigation and useful results."

In a late number of the *Country Gentleman*, Mr. R. T. Selden, of Westchester Co., New York, has reported some experiments in feeding, that seem to throw a clear light on this subject. "I have been trying," he remarks, "some experiments with corn-stalks, and also with corn and cob meal. When the stalks are cut about an inch long, I find my cattle and sheep will eat nearly one-third more of them than of the whole stalks. Also I find that when I cut them much smaller—say one-fourth or one-eighth inch—and pour hot water over them, to stand a few hours, the difference is so great I can hardly believe it. They do not leave a gill, and they thrive on it wonderfully. I find, by a fair and careful trial, that they gain on this fodder more than on the best hay.

"I have also tried grinding my corn in the ear with an iron mill. I have long thought the cob to be nutritious, and I have now given it a full trial. I believe it is said the cob weighs one-fifth of the whole ear. For the sake of giving it a fair test, I have fed alongside of this another preparation, which is corn-meal without the cob, but instead of the cob one-fifth weight of oat-straw cut fine. In this mixture the weight of oat-straw is the same as the weight of the cob in the other, so as to make a fair comparison.

" After trying for several weeks, I found the corn and cob meal came out a little ahead; that is, the cattle and sheep fed on the corn and cob meal gained more in weight than those fed with meal and cut straw. I do not know how much value there is in oat-straw; but, so far as my experience goes, I think the corn-cob a little the best. I have also tried this corn and cob feed mixed with cut stalks, and when the mass is well soaked in hot water, it makes one of the best feeds I have ever used. One of my neighbors says he thinks this mixture almost equal to pure grain for fattening beef or making butter. It seems to me that if these three parts of corn are made fine and mixed in the right quantity, it must make a most excellent kind of feed for nearly all purposes.

" All the above experiments were carefully made by weight and measure."

But there is another consideration commending to farmers the use of this feed. By employing a portable mill or crusher, of which there are several kinds in successful use, they will find both economy and convenience in grinding their ears at home, instead of shelling the corn and sending it to the mill. It has been estimated that the expense of shelling the corn, conveying it to the mill, paying the toll, and transporting the meal back to the farm again, is sufficient to pay for crushing twice the quantity at home. That is, twenty bushels of the grain alone, ground at the mill, would cost as much as forty bushels of grain, with the included cob, ground at home. In the opinion of some, the gain is even more than this.

According to this estimate, the farmer who uses a crushing-mill not only gets his cobs ground for nothing, but he also gets his grain ground at half price.

On the other hand, it is the opinion of some feeders that this mode of grinding does not render the meal sufficiently fine to get the full benefit of the nutritive value in feeding. There is, perhaps, some weight in this objection if the meal is fed raw or dry. But if cooked or steamed before feeding, or even if thoroughly soaked, the objection is obviated, and a still further value is imparted to the feed.

An Ohio correspondent of the *Country Gentleman*, after an experience of five years in grinding his ears at home, pronounces it a complete success. He converts, as he informs the editor of that journal, from six to ten bushels of ears per hour into meal, which he feeds to chickens of two days old, and so on, up to the ox of two thousand lbs. ; and fattens from eight to twelve head of cattle every winter with satisfactory returns.

A correspondent of the *Prairie Farmer*, after feeding his cows for over two years by this method, states that they do much better than formerly, and that he finds a gain of one-third in the use of this feed.

If some definite conclusion, or even a probable estimate were formed, as to the relative value of the cob, *when compared with any known standard*, it would give perhaps a clearer view of what it is actually worth in feeding. Chemistry has shown that seventy-three parts in two hundred are more or less nutritious. Experience has also proved that it possesses this quality in a considerable degree. But here opinions differ, some rating its nutritive value much higher than others.

By most farmers who make use of the cob, it is compared with the straw of the cereals, and seems to be considered equal to them on a general average. The editor of the *Rural Annual* considers it equal to good wheat-straw. Some others compute its value to be quite equal to that of the best oat-straw. If we take the average value of the straw of wheat, oats, barley, and rye, compared with hay as a standard, it will give three hundred and fifty pounds of the former, equal to one hundred pounds of the latter. That is to say, the value of good hay is three and a half times greater than the mean value of those straws. If this ratio is taken to represent the value of the corn-cob, as compared with that of hay, the estimate would seem to be, at least, a reasonable approximation to the truth.

NUTRITIVE VALUE OF CORN AND COB MEAL.—If we now calculate the value of the grain, by referring it to the same standard, we shall then be able to see how the cob compares in value with the corn, and also to determine the value of the corn and cob meal, as compared with that of hay.

The grain of Indian corn has been variously rated as to its actual worth for feeding. This must necessarily depend, in some measure, upon the animal to which it is fed, and in part upon the object for which it is given. Its general nutritive value, according to Prof. Johnston, as indicated by experiments made by different persons in different countries, is to that of hay as one to two ; five pounds of it being given as equal to ten pounds of hay. But when the object is to produce beef, butter, mutton, or pork, its effective

9

value is very much greater. Three pounds of corn, and even less, have not unfrequently produced a pound of pork; and for making mutton, it is said to be even more effective than this.

On the whole, the feeding value of this grain, when used for the purpose of converting it into any of the above products, will perhaps be fairly represented by taking forty pounds of it, and when used for other purposes, fifty pounds, as equivalent to one hundred pounds of hay. This will give a mean ratio of forty-five pounds of corn to represent one hundred pounds of hay. But to render the estimate free from any reasonable doubt, let us take forty-eight pounds as equivalent to one hundred pounds of hay; then, according to the valuation given above for the cob, we shall have:

Lbs. of Hay.		Lbs. of Corn.		Lbs. of Cob.
100	=	48	=	350

There are, therefore, in one thousand lbs. of corn and cob meal:

	Lbs.		Lbs.
Ground Corn..................800	=	1,666 of hay.
Ground Cob...................200	=	57 "
Corn and Cob Meal........1.000	=	1,723

From this comparison it appears that the value of corn and cob meal is seventy-two per cent. greater than that of hay, and it is by no means improbable that further experience, and more systematic experiments in feeding, will show, what some already believe, that its true value is higher than this.

Philosophy teaches us that things, apparently trivial in themselves, sometimes derive from circumstances a consequence unperceived by the casual observer. Some idea associated with quantity or numbers, some relation to a system or class, some fact illustrating an undeveloped possibility; these, or similar causes, frequently rescue from insignificance an object deemed useless or paltry, and invest it with an unsuspected interest, dignity, or value.

The farmer who casts out the corn-cob from his crib, as a thing utterly worthless, and fit only to be trodden to the earth, is probably unconscious of the utility it is capable of, and still less aware of the extent and value of the class it represents; little suspecting that the corn-cobs raised every year in the United States, contain a sufficient amount of nutriment to winter seven hundred and fifty thousand cattle, and are worth in the aggregate not less than fifteen million dollars.

NUTRITIVE VALUE OF CORN AND COB MEAL COMBINED WITH CHAFFED STALKS.—Let us now examine the economical value of the feed we have been considering, when further combined, as it frequently is, with the stover of corn finely chaffed. Let us take the three several products of the corn crop—the grain, the cob, and the stalk—and suppose them to be combined in the same proportions in which Nature produces them.

We will take, for illustration, the product of one acre, assuming the yield to be one hundred bushels of grain, at fifty-six pounds to the bushel. The proportion of stalks will be about one ton to each twenty-five bushels of grain, and the weight of the cob is one-quarter of the weight of the grain. We shall therefore have, as the product of the acre, the following amount of feed:

Grain....................................5,600 lbs.
Cob...1,400
                                          ____
Corn and Cob Meal.....................7,000
Stalks......................................8,000
                                          ____
Total product of the acre.........15,000

Now by comparing these with the same standard of value as before, the stover being equal to its weight of hay, and the meal being equivalent to seventy-two per cent. more than its own weight of hay, we shall find the feeding value of the acre of corn to be as follows:

	Lbs.		Lbs.
Corn and Cob Meal................	7,000	=	12,040 of hay.
Chopped stalks....................	8,000	=	8,000
Total amount of feed..........	15,000	=	20,040

With some varieties of corn, however, the proportion of stover would be less than the above estimate. Taking the product of stalks, for such cases, at three tons instead of four, the total amount of feed would be thirteen thousand pounds—equal to eighteen thousand and forty pounds of hay. In one case, then, we have nine tons of hay, and in the other case ten tons, as the measure of the feeding value of one acre of corn.

In every instance where this feed is employed for the purpose of being converted into beef, mutton, butter, etc., the effective value of the grain is greater than we have estimated it, and will give a higher result. If the entire product of the above acre were used for fattening purposes, the aggregate feeding value would be from two to three tons higher than the estimate here given.

It will perhaps be said by some, that one hundred bushels of corn per acre is an uncommon yield, quite out of the ordinary range, and that most farmers therefore would not be able to realize the above result. From this opinion, or rather from the inference in the last clause, we must beg leave to differ. We do not believe that there is a farmer in the United States who is *not able* to raise one hundred bushels of corn per acre. He may lack the *knowledge*, or the *resolute purpose*, and there are many who lack both. But where these are both present, the ability is not wanting.

The farmer who makes up his mind to raise one hundred bushels of corn on an acre, will generally do it. He who begins by saying he is " not able," will certainly not do it. He will take good care to keep his word. Such men are usually very tender of their veracity. Those who are continually proclaiming to the world, when any thing comes up to be done, that they are " not able " to do it, have invariably one advantage. They require no logic to prove their assertion. The world is always ready to believe them.

"Not able" is indeed a mischievous phrase, that has done much harm in society as well as in agriculture. The peculiarity of the evil is, that the want of faith produces the inability. The assertion, in a certain sense, creates the fact, and makes that virtually true which was not true before. This phrase is the enemy of the farmer, and should be pursued and exterminated with the same zeal that he employs in pursuing the vermin that devour his crops, or the weeds that infest his soil.

We repeat, then, that every farmer who chooses may raise one hundred bushels of corn, and even more than this, upon an acre. How much better he can do than this must depend upon himself. As this question, however, is considered in a subsequent chapter, it need not here be dwelt upon.

Let us now compare the nutritive value of corn per acre with that of some other leading crops, taking the yield of the latter on the same scale as the above yield of corn. It will probably be conceded that sixty bushels of wheat, ninety bushels of oats, fifteen tons of potatoes, and twenty tons, on an average, for other root crops, per acre, would be at least as large a yield in proportion for these crops as one hundred bushels would be for corn.

By referring the nutritive value of these several products to the same standard as before, which was good meadow hay, it will be found that the wheat crop, as compared with corn, yields about one-half the amount of nutritive value per acre; the oat crop still less than the wheat; the potato about three-.

fourths the amount of corn, and the other root crops less than the potato.

To state the comparison more definitely, it would be very nearly as follows:

Nutritive value of Wheat, per acre = 10,000 lbs. of hay.
   "   "   Oats,   "   = 9,000
   "   "   Potatoes "   = 15,000
   "   "   other root crops = 14,000
   "   "   Corn, per acre = 20,000

COST OF PRODUCTION.

This is a subject of vital interest to the cultivator. It is indeed the great practical question in husbandry, to which all others are justly considered subordinate. When the farmer has harvested his crop, it is his first concern to know what every bushel of grain has cost him. Whether his yield is fifty bushels per acre, or one hundred and fifty, is doubtless a matter of some consequence, and one which he is likely to understand; but whether it has cost him twenty-five cents a bushel, or seventy-five cents, is a matter of still higher moment, and a question far too important to be settled upon any principle of guess-work, as it too frequently is.

To be accurately posted on this point is neither impossible nor difficult, and is moreover quite indispensable to the success of the farmer as a business man. Though a large yield of corn, nay, even the largest yield of his town or county, is to every enterprising cultivator an object of commendable ambition; yet, when the object is achieved, the value to him of such yield depends, after all, upon what it has cost him, and not merely this, but also upon his *knowing the cost*. In fact, without this knowledge, all the business of his farm is at loose ends, and all the operations he embarks in are involved in uncertainty.

It is easy to perceive that the cost of producing this crop must necessarily exhibit a marked diversity in different sections of the country, and this is especially noticeable in comparing the expenses of corn-culture at the East with those of the West. But there are also other causes of difference, so numerous and pervading, that it is a rare circumstance to find any two estimates in all respects alike, even in the same section or neighborhood.

For the purpose of comparison and reference, we here submit a few statements and estimates relating to the expense of corn-culture in the various sections of the country that have been reported during the past two decades.

In the Transactions of the New York State Agricultural Society for 1848, Levi T. Marshall, of Oneida County, is reported to have raised eighty-seven bushels per acre, on two acres of ground, at a cost of seventeen and three-fourths cents per bushel, making the expense per acre equal to fifteen dollars and forty-four cents.

A farmer in St. Lawrence County, N. Y., at a later date, reports the cost of his crop to the *Country Gentleman* at twenty-seven cents per bushel, and sixteen dollars and sixty-six cents per acre.

Another farmer, of Shelburne, Mass., in a communication to the same journal, states that the cost of his crop was twenty-three cents per bushel for the grain, being at the rate of twelve dollars and thirty-six cents per acre.

Mr. Dickerman, of Conn., reports to the *Agriculturist* a crop of one thousand bushels, raised at an expense of about six hundred dollars, which is sixty cents per bushel, the expense per acre being about twenty-six dollars.

H. S. Senter, of Mercer County, Ill., writes to the same journal that his crop of one thousand four hundred and forty bushels was raised at an expense of ninety-one dollars, which is less than seven cents per bushel.

Mr. Walker, of Concord, N. H., gives the expense of a crop raised by himself at forty-nine and three-fifths cents per bushel, or twenty-four dollars and eighty cents per acre.

Jonathan Roberts, of Montgomery County, Penn., has calculated the expense of his maize crop at nineteen dollars per acre, and thirty-one and two-thirds cents per bushel.

In Mr. Colman's Report to the Legislature of Massachusetts he gives two crops from the same town, showing a very wide difference in the expense of raising them; the one costing nineteen cents per bushel, and the other fifty-seven cents. In this case one farmer paid three times as much per bushel for his corn as the other.

A crop raised in Deerfield is quoted in the same report as costing twenty-two dollars and sixty-seven cents per acre, and forty-five and one-third cents per bushel; and another crop in Shelburne is given at thirty-five dollars and seventy-seven cents per acre, and fifty-one cents per bushel.

A correspondent of the *Prairie Farmer* reports a crop in Warren County, Ill., of four thousand bushels, that cost from nine to ten cents per bushel of ears.

In a report recently made to the Whately and Deerfield Farmers' Club, Mass., Edward C. Parker is stated to have produced a crop of corn at a cost of forty-two and a half cents per bushel. In this case the net *profit per acre* was eighty-three dollars and forty-four cents, the grain being estimated at one dollar per bushel, and the stover at ten dollars per ton.

In the same report, the crop of Charles Hagar is given at a cost of forty-three and a half cents per bushel. The profit per acre was here fifty dollars; the price of the corn and stover being the same as above. These were prize crops, the former taking the first and the latter the second premium.*

In the statements here presented the expense per acre, as far as given, averages twenty-one dollars and fifty-eight cents. The cost per bushel ranges from seven cents to sixty cents, giving an average of thirty-two cents.

---

* In these two cases the cost of each crop was calculated without deducting the value of the stalks. But the latter is the method of estimating more generally practised. Had the value of the stalks been here deducted from the expense, the cost per bushel would have been very much less. In the former case it would have been twenty-four cents per bushel, and in the latter nineteen cents.

In one of the Annual Reports of the Patent Office is a series of statements from farmers in nearly all the States of the Union, in which the estimated cost per bushel ranged from seven cents in Iowa to seventy-five cents in Massachusetts, making an average of about twenty-seven cents per bushel. If we combine this with the above average of thirty-two cents, it will make a general average of about thirty cents per bushel, which is probably not far from the true cost of production for Indian corn in the United States during a period extending over the last twenty years.

The extreme figures between which this average lies are seven cents in Illinois and Iowa, and seventy-five cents in Massachusetts. Probably the actual difference between the two sections of the country would be fairly stated if we should call the average cost of production in the Western States fifteen to twenty cents per bushel, and in the Eastern States thirty-five cents.

On the other hand, this marked difference between the East and the West, in regard to the expense of raising corn, is perhaps compensated, if not more than this, by the difference in the market value of the grain.

But let us look a little more closely into this question of cost of production, to discover whether it is possible to reduce it below the present average, and if so, by what means. A careful investigation of the subject will perhaps make it appear that the true method of reducing the cost per bushel of Indian corn is to be found in increasing rather than diminishing the expense per acre, provided this is done on sound principles, and with good judgment. In other words, the farmer who adopts the best methods of culture will discover that, as a general rule, and up to a certain limit, the more he pays out per acre for extra culture and fertilization, the more grain he will get back, and the less will be the cost per bushel.

There is a certain amount of work that *must* be done upon each acre of ground, before any grain whatever can be produced. A certain amount of expense is inevitable for even the lowest rate of production. Let us, then, endeavor to ascertain this lowest limit of expense per acre.

No farmer attempts to raise a crop of maize without ploughing the ground, at least once. His land has then to be marked out and planted, and when the grain is ripe, the crop is harvested and stored. This may be considered the lowest stage of corn-culture in which there is no manure employed and no after-tillage. The items of expense, in this case, would probably be, on an average, as follows:

Ploughing	$2 00
Marking, planting, and seed	2 00
Harvesting	3 00
Rent	5 00
	$12 00

These figures are of course variable, according to locality and other circumstances, but will be found on an average very nearly as stated.

The yield of the above acre must, of necessity, be very low, and cannot safely be estimated at more than fifteen bushels. The stalks, according to the ratio before given, would amount to three-fifths of a ton, which, at six dollars * per ton, would be three dollars and sixty cents. We should then have the following result:

<pre>
Total expense of one acre.................$12 00
Deduct value of stalks.................... 3 60
                                          ────────
Cost of fifteen bushels of Grain........ $8 40
</pre>

which makes the cost per bushel fifty-six cents.

This estimate includes only those items of expense that are *unavoidable*. The other usual expenses of corn-culture, including harrowing, cross-ploughing, after-tillage, manure, etc., are all *optional*. Now here, in these optional expenses, is precisely where the profit lies.

The farmer has in this case paid out twelve dollars to bring his acre up to the point where production begins. After that, every dollar judiciously added goes straight to the mark, and tells powerfully on the yield. The outlay of twelve dollars he is compelled to incur, before he realizes a single kernel, and whether he gets five bushels or fifty. The other expenses are discretionary and variable, and what is most important the crop varies with them, and can only be increased by increasing them, or some part of them.

---

* This is much too low for the true value of good stalks; but as there are always some farmers who insist on computing their stover at half its real worth, it is perhaps as well to adapt the illustration to their standard.

To illustrate this, let us take the same acre, and superadd the following treatment: Let the ground be cross-ploughed and harrowed before planting, and let the crop be twice cultivated during its growth. The whole expense would then foot up as follows, allowing for a slight addition to the cost of harvesting :

Ploughing twice and harrowing..........	$4 50
Marking, planting, and seed..............	2 00
After-culture, twice through.............	4 00
Harvesting...............................	3 50
Rent....................................	5 00
	$19 00

The yield in this instance, with ordinary care, would probably reach thirty-five or forty bushels; but may safely be assumed at thirty bushels. The stalks would, in that case, amount to one and one-fifth tons, which, at the value above stated, would be worth seven dollars and twenty cents, giving the following result :

Total expense of crop..................$19 00
Deduct value of stalks.................... 7 20
Cost of 30 bushels of grain...........$11 80

bringing the cost per bushel to thirty-nine cents and a fraction.

We will now suppose this acre to be cultivated in a more expensive manner, by increasing the amount of tillage and adding manure, making the entire cost as follows :

```
Previous expense........................$19 00
To which add:
  Manure...........................$16 00 ⎫
  Subsoiling........................  3 00 ⎪
  Extra harrowing....... ..........  1 00 ⎬  20 50
  Increased expenses of harvesting....    50 ⎭  ______
        Total expense of crop.................$39 50
```

Taking the yield of this crop at an average probability, it could scarcely be less than seventy-five or eighty bushels. But calling it seventy bushels, this would give two and four-fifth tons of stover, worth sixteen dollars and eighty cents.

From the above expense is to be deducted, not merely the value of the stalks, but also one-half the outlay for manure and for subsoiling, as the effect of these is not limited to a single season, but extends to successive crops. The net result, therefore, of this crop will be as follows:

```
Total expense...............................$39 50
From which deduct half the cost of manure and
    sub-soiling................................  9 50
                                                ______
                                                $30 00
From this deduct value of stalks...............  16 80
                                                ______
    Cost of 70 bushels of grain..............$13 20
```

which makes the cost per bushel about nineteen cents.

Now, on comparing the results of these three crops from the same acre, we find that by increasing the expense of cultivation, the cost per bushel is reduced successively from fifty-six cents to thirty-nine cents and nineteen cents. This reasoning is of course theoretical, but if the figures assumed are reasonable and consistent with experience, it carries with it the force of great probability.

By extending the illustration still further, we should find that the cost of production would be reduced yet lower. If, for instance, in the crop last given, the amount of manure were doubled, and the after-culture again repeated, the probable yield on a fair soil with good management would not be less than one hundred bushels per acre, which would bring the cost per bushel to sixteen cents.

On comparing the figures assumed throughout this investigation with those of similar crops actually raised and frequently reported, it will be found that the estimates above made are entirely probable, and that any reasoning founded upon them becomes a fair presumptive argument.

In the following table the results of these four successive crops are brought together, omitting fractions, for the purpose of comparison. The table also indicates the money value of each crop, supposing the corn to be worth seventy-five cents per bushel, and the stalks six dollars per ton, and it further shows the rate of profit per acre, per bushel, and on the investment.

In the second column the net expense per acre is obtained by deducting the value of the stalks from the gross expense. This mode of estimating the cost of corn is usually adopted by farmers, though not strictly correct.

In the fifth column the money value per acre includes the value of the grain and stalks:

CROP.	1 Total expense per acre.	2 Net expense per acre.	3 Cost per bush.	4 Prod. per acre, in bushels.	5 Money value per acre.	6 Profit per acre.	7 Profit per bush.	8 Profit on the Investment.
1st .	$12 00	$8 40	56 cts.	15	$14 85	$2 85	19 cts.	28 per ct.
2d..	19 00	11 80	89	80	29 70	10 70	86	56 "
3d..	80 00	18 20	19	70	69 80	89 80	56	181 "
4th.	40 00	18 00	16	100	99 00	59 00	59	147 "

This table is based on plain and simple principles of husbandry. But by the use of special manures, by more elaborate disintegration of the soil, and by closer planting, which last is only warranted with copious manuring and deep pulverization, a still lower cost of production might undoubtedly be reached.

There is a large class of soils in which all the elements of maize are to be found, with the exception of some one or two that happen to be almost entirely absent. In every such instance the application of special manures, if rightly selected, is attended with the highest advantage, not unfrequently doubling the yield at a trifling expense.

But in order to obtain such results, the soil must be understood by its owner. Unless he knows precisely which element is wanting, he is very unlikely to be successful in supplying it. If he applies the wrong fertilizer, he might nearly as well apply none. In either case, the crop will hardly be worth gathering. But if he gives to his land the specific manure called for, it not merely increases the product, but it makes all the difference between a maximum and a minimum yield. With the wrong fertilizer or with none, the cost of the grain would very likely be a dollar per bushel, or more; with the right fertilizer it would probably be ten cents per bushel, or less.

. In all such cases as this, and also in every instance where the soil is naturally and unusually rich in corn-elements, and in nearly all cases of the highest and best system of culture, the cost of production could probably be brought to a lower figure than the lowest in the table.

But there are other points in the table that deserve attention. By referring to the second and fourth columns it will be seen that in the first crop the farmer gets fifteen bushels of corn at a net cost of eight dollars and forty cents. In the second crop he gets the same, and fifteen bushels more for an additional cost of three dollars and forty cents; showing that the second fifteen bushels cost him less than half the price of the first. In the third crop he gets the first thirty bushels at the same cost as in the second, and forty bushels more at an additional cost of one dollar and forty cents. In the fourth crop the first seventy bushels cost him the same as in the third, and he gets thirty bushels more for an additional cost of two dollars and eighty cents.

It is also worth while to notice the ratio of increase in the profit per acre, as compared with the amount invested. Thus when in the first crop he invests twelve dollars, the profit on the acre is but two dollars and eighty-five cents. In the second crop, by adding seven dollars to the investment, the profit per acre rises to ten dollars and seventy cents; showing that while the first twelve dollars earn twenty-three per cent., the next seven dollars earn over one hundred per cent. In the third crop, by adding eleven dollars more to the investment, the profit on the acre reaches thirty-nine dollars and thirty cents; and finally, by adding, in the fourth crop, ten dollars more, the profit per acre rises to fifty-nine dollars.

It will also be observed, by referring to the eighth column, that the rate of profit on the capital employed advances by a ratio no less rapid and remarkable from twenty-three per cent. to one hundred and forty-seven per cent.

Now it is not claimed that the illustration here presented has all the precision and certainty of a mathematical demonstration. Yet it is believed to be a fair statement of average results such as would occur in the ordinary practice of farmers. In assuming one hundred bushels as the yield of the fourth crop, the amount is doubtless liable to exceptions. It would not probably be reached in an adverse season, nor on a poor soil, and least of all by a slovenly farmer who is wiser than all the books and journals. But it is a product often obtained at a *less outlay* than the amount assumed, and it will scarcely fail to be equalled or surpassed, when this amount of expense is rightly applied.

But there is another contingency that is liable to affect some of the figures in the above table, and which would render them much more striking, though no less correct. When corn is consumed on the farm where it grows, it pays the owner a better price than the market quotations. Very many farmers, by converting this grain into pork, mutton, beef, or butter, are enabled to realize for it a dollar or more per bushel, even when it is bringing seventy-five cents or less in market.

Now, if the price of corn were taken at one dollar in the table instead of seventy-five cents, the results, or a part of them, would be not only more remarkable, but, in a large class of cases, nearer the truth. The third and fourth crops, at this price, would give the following exhibit :

CROP.	Total expense per acre.	Net expense per acre.	Cost per bush.	Prod. per acre in bushels.	Money value per acre.	Profit per acre.	Profit per bush.	Profit on the investment.
3d..	$30 00	$13 20	19 cts.	70	$86 80	$56 80	81 cts.	189 per ct.
4th.	40 00	16 00	16	100	124 00	84 00	84	210 "

Perhaps the most instructive lesson contained in these tables is to be found in the great principle which stands out clear and conspicuous, that the last part of the yield, or *the extra yield produced by each addition to the expense, is the part that pays the profit.*

This principle is well understood and acted upon in some other branches of industry, and why should it not be equally improved in husbandry ? Publishers have long since discovered that the success of a book or magazine depends mainly on the excess of sales beyond a certain number of copies, and their plans are shaped accordingly.

Journalists are well aware that the profit of their business lies in the last ten, or twenty, or fifty thousand of their circulation. Hence any and all means by which this vast circulation may be secured are promptly and fearlessly adopted, without regard to expense. Ingenious and costly expedients are employed to swell the subscription list, and the reading community are reached, through every channel of approach, with a splendid array of inducements to subscribers. The catalogue of liberal offers embraces an endless variety of things useful and ornamental which are equally creditable to the taste, skill, and munificence of the proprietor. To furnish out the list of premiums, the world of art, and two of the kingdoms of nature, are laid under contribution. Seed-packages, gold pencils, valuable barometers, and costly engravings are marshalled into the service to enact the part of canvassers.

Some things hitherto applied, through ignorance, to mistaken uses, have by this means been restored to their proper functions. It has thus been discovered that the true and original design of the strawberry is to attract subscribers to the better class of journals; and that sewing-machines, formerly supposed to be useful in constructing garments, were mainly intended by the inventor to guide the popular choice in selecting the best periodicals.

All this is doubtless right and proper in a business

way—a strictly legitimate proceeding, which proves the sagacity of the publisher, who clearly comprehends, not merely the absolute necessity of the first ten thousand subscribers, but the gilt-edged value of the last fifty thousand.

In like manner, and on the same principle, the clear-headed and well-informed cultivator will be prompt to perceive and appreciate, not only the utility of the first twenty or thirty bushels of his corn crop, but also and equally the gold-bearing value of the last fifty or one hundred bushels which are added to the yield by a slight increase in the expense of culture.

To secure the latter yield, the farmer need not resort to any costly means of tempting Nature into an abnormal munificence. She is never wanting in generosity to those who are in true sympathy with her, who study out her laws and comply with them.

Instead of large disbursements for premium lists, the farmer has only to invest his spare dollars in improved implements, fertilizing materials, and agricultural journals. These are the great agents and insurers of successful husbandry, and no cultivator of the soil who understands his interest will ever be without them.

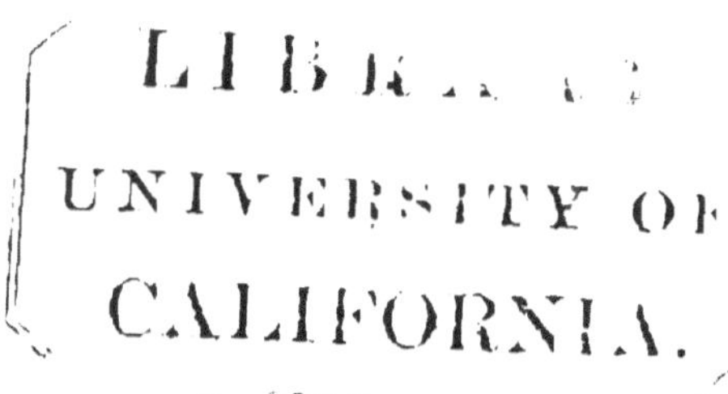

HOW TO OBTAIN A LARGE YIELD OF CORN.

Having treated of the productive capacity of Indian corn; having shown by numerous instances, as well as by reference to natural laws, that it is capable of a more bountiful yield than is usually obtained; and having urged upon farmers the possibility as well as the duty of increasing the acreable product of this grain, they will perhaps turn to the author with the natural inquiry, "How are these desirable yields to be achieved?"

If we refer them to other pages of this book where general instructions on the subject are laid down, there are some who will, perhaps, insist that the information given is not sufficiently definite, and will ask for more specific directions; considering it a matter of course that a work devoted to the subject should furnish a perfect form for each particular process, and some infallible fertilizer equally adapted to every variety of soil.

To those who entertain this view of the matter, a book that should prescribe an exact method of raising corn to be mechanically followed in every case from beginning to end, laying down in detail each particular step to be taken without discretion, proposing in connection with this a specific manure to be used on all occasions, and claiming that the infallible result would be a marvellous and unheard-of yield, would doubtless prove attractive, and probably be hailed as a useful work.   But the author of such a book would justly be pronounced a charlatan, and would deserve their contempt.

There are, probably, very few farmers in the country who do not know that there is not, and cannot be any patent, labor-saving process for turning out two or three hundred bushels of corn from an acre.   The mistake of these men lies in expecting too much. They would have the results without complying with the conditions.   They do not seem to remember that large yields of corn, as of other crops, have never yet been stereotyped, to be mechanically reproduced at will; nor do they, on the other hand, spring from accident or neglect.   They are the prizes which industry, intelligence, and skill carry away in the face of contingencies, and in spite of obstacles.

If there were indeed a new, easy, and infallible method of raising corn, with all the elements of uncertainty left out, promising large yields with little labor and less thought, and imposing no tax upon either muscle or brain, we should then see all the world turning farmers, and all the farmers growing rapidly rich.

But, happily, the class of men who indulge these

absurd expectations are few in number. The great majority of American agriculturists are men of sense and reason. When they aim at large returns, they expect to make corresponding exertion. When they inquire how they are to obtain a large yield, they do not imagine they can dispense with the needed effort. What they wish to know is, how to secure a large product with reasonable effort, and at the same time make the yield a *profitable* one. This is the inquiry continually and everywhere raised by intelligent farmers.

Although a *general* answer to this inquiry is to be found on other pages of the present work, yet, in view of the importance of the subject, and for the sake of a more full illustration, we will endeavor to present an answer, if possible, more definite, clear, and distinct.

In every stage of corn culture, from the first to the last, there is some one method that is better than any other. In the preparation of the ground there are many different ways of proceeding, but there is only one best way. For enriching the soil there is a numerous catalogue of manures, and various modes of applying them; but some one, or a few of them, are better than all the rest. The same is true in regard to steeping the seed. Also in the distribution of the grains at planting, as well as in the depth of covering them, there is for each a diversity of plans; but there is one that will give the largest yield. In like manner, the after-culture also has its own superior process that is more productive than any other.

When all these preferable modes are ascertained

by the farmer, and blended together in one system of culture, they form a comprehensive BEST METHOD, which embraces all the conditions of success, and must, from the very nature of the case, give higher results than any other. But this method is *different for each different soil,* and varies according to other varying circumstances.

Yet for every farmer in the country, without any exception, there is such a method. By means of experimental processes, elsewhere explained, he can determine these several conditions of success, and in adopting them, he adopts the true and the only sure method. If he will take the time and make the effort necessary for this purpose, he may acquire from such a series of experiments that knowledge which no other man can impart to him, and which will enable him to obtain, beyond any question, a maximum yield of corn at a minimum cost per bushel.

Let us now, for illustration, suppose the case of a farmer who has during the past year introduced into his corn crop a sufficient number and variety of such experiments to determine all the points that he needs to know for insuring the success of his next crop. He has ascertained what are, in his case and for his soil, the most certain and productive processes, and is therefore prepared to lay out his plan for the ensuing year.

If you interrogate him on the subject, he speaks with confidence, and not with the tone of a man who is guessing, or groping in the dark. If you ask him how he is going to proceed, and his reason for each process, his answers are not vague and hesitating, but prompt and definite. He has worked out his *best method* with hand and brain, and realizes that he is master of the situation. We will now suppose him to describe, as follows, his plan of operations for the year to come, as deduced from his experience in the year past :

"In preparing the ground for my next year's corn crop," is his language, "I shall commence *in the fall* by ploughing to the depth of ten inches, following the surface-plough with the subsoiler as deeply as it can be made to go. That ten inches is the right depth for my soil I have clearly proved, and the utility of subsoiling was shown last season by its increasing the yield more than thirty per cent.

" After the ground is thus ploughed, I shall apply a moderate top-dressing, fifty bushels to the acre, of lime and unleached ashes mixed in equal proportions. After having brought my land to this stage of preparation in the fall, I shall then resign it for the winter season to those ever-useful and unfailing friends of the husbandman, the frost and rain, the storm, the sunshine, and the brooding snow. While the farmer wears away the hours of winter in recreation or repose, these friendly agents will do his work with tireless industry, subduing and meliorating the soil with a magic effect ' beyond the reach of art,' and their useful work will be all the better done, and the more effectual, in consequence of the aid previously rendered by the plough and the subsoiler.

" In the spring, before planting, I shall again apply the surface-plough one inch deeper than before, to be followed again with the subsoiler. I shall next apply twenty loads to the acre of stable manure, previously sprinkled with brine. This application is to be immediately turned under with a shallow plough, and thoroughly harrowed into the soil. My seed before planting will be steeped twenty-four hours in a weak solution of hen manure and chloride of lime. I shall then plant my corn, which is the King Philip variety, in drills three feet apart, with the grains eight inches asunder, covering the corn to the depth of two inches.

"In the drills along with the grain is to be applied a fertilizer, prepared by adding together two parts of leached ashes, three parts of the flour of bone, and five parts of well-rotted stable manure. After the corn is fairly up, I shall apply a moderate dressing of ashes, plaster, and lime. For the after-culture I shall use the horse-hoe and hand-cultivator often enough to keep the soil well aërated, and to prevent the growth of weeds. As the ground before planting will be thoroughly pulverized, the plough will not be required in the after-culture.

"This is the course that has been clearly pointed out to me by my experience of the past year. *All the details of this plan were indicated and proved to be the best by the results of my experimental crop.* I therefore accept the plan without hesitation. I do not know that it would be the best method for my neighbor, with a different soil, but I am sure that it is the best for me. In examining and comparing the results of my experiments last fall, I found that in that portion of the field *where all these conditions met together*, the yield was equal to one hundred and seventy bushels per acre of grain, and the stover was at the rate of over five tons, surpassing, by nearly thirty per cent., any other results of the experimental crop.

"Now, seeing that I shall adopt for the coming year, the precise method that gave this yield, applying it also to the same soil, thus complying with all the conditions exacted by the laws of Nature, I consider that, with a propitious season, there is a fair probability of reaching the same amount again. If, however, I succeed in getting one hundred and fifty bushels, I shall be well paid and well satisfied.

"I have ascertained, by experiments in feeding, that there is a way to convert my corn-stalks into milk, cheese, butter, and beef, so as to realize for them over ten dollars a ton; and my corn, by the same best method of feeding, can be turned into pork or beef, so as to bring on an average over one dollar a bushel. I find, on computing the expense per acre of this method, that it amounts to fifty dollars, after deducting one-half the cost of the manure, and one-half the expense of subsoiling, both to be charged to future crops."

This is supposed to be the language of a farmer applying the knowledge deduced from his experiments, to guide him in his future operations. There is in his statements nothing unreasonable, extravagant, or impossible. The case assumed not only illustrates the general principle, but fairly represents the results that would probably be derived in a majority of instances from pursuing a system of best methods.

But let us now compute the cost per bushel of this crop. The yield would probably be one hundred and fifty bushels per acre, as estimated. It might be more, it might be less. To meet varying possibilities, we will take several estimates, viz.: one hundred and twenty-five bushels, one hundred and fifty bushels, and one hundred and sixty-five bushels. The product of stalks for these yields would be from five tons upward. Taking the lowest rate, it would be five tons. The amount assumed as the expense of the crop, fifty dollars, is probably what it would be on an average. We should then have the following result:

```
Cost of crop...................................$50 00
Deduct 5 tons of stalks (at $6 per ton)......... 30 00
                                               ─────────
                                                $20 00
```

This would make the cost of production, omitting fractions :

```
For a yield of 125 bushels, equal to....16c. per bushel.
For a yield of 150 bushels,   =    .....13c. per bushel.
For a yield of 165 bushels,   =    .....12c. per bushel.
```

There may be those who will differ in opinion as to this estimate. One will perhaps say that the quantity of manure is not sufficient to account for such a product; another, that the planting is too close to give such a yield of grain, etc. But certainly a method, comprising half a score of different processes and conditions, cannot fairly be judged by any one of them considered separately. The yield is not the result of any one element in the plan, but the joint product of the whole of them. Moreover, each part of the plan is adapted to all the other parts, and the whole together are adapted to a particular soil. The method and the soil are the counterparts of each other, and the highest capability of each is only developed when they are brought together.

The above method with a different soil might be entirely inadequate to such a yield—might even be a total failure; and the soil, whatever its merit in this connection, might, under a different treatment, give a very uncertain result. But when the method and the soil are perfectly fitted to each other like dove-tailed joinery, or like the wheels and grooves of machinery, it would seem as if results larger than these might easily follow.

When the conductor of an electrical machine is highly charged, you may apply a hundred different objects to it, and none of them is qualified to extract the fluid. But there are substances in nature precisely fitted to produce this effect, and the moment you apply one of these the spark flies, and the fluid is given up. So when the farmer applies to his soil the method that is exactly suited and congenial, it surrenders its prolific virtue with an exuberance before latent and unsuspected.

Now there are some soils which, without any manure whatever, are capable of yielding, and with deep and thorough culture have yielded, over one hundred bushels per acre. It seems, then, reasonable to infer that if such a method as the above is applied to the soil to which it is precisely adapted, to the very soil, in fact, out of which it is created, it can scarcely produce less than the amount assumed, and would very probably yield more.

But there are many cases in which the result would be more striking than that above given. One farmer would find his soil so constituted that a less amount of tillage than the above would give an equal effect, or the same amount would give a larger effect. Another would find his land so prolific by nature that the amount of fertilization stated above would give a larger product than that assigned to it. A third would find that his soil lacked only one or two elements of fertility, while containing all the rest in ample abundance, and that these absent elements could be added at less than half the expense of the above manure. In such cases as these, if all the other conditions of success were complied with, the cost of production would be found lower than in the above estimate, and the yield larger.

Thus, by a system of unsparing investigation, each man perfects his own method, and acquires for himself all the knowledge essential to the highest success. He may derive valuable aid from books and journals, from tables and formulas, from chemical analysis, and from all the countless instances of recorded experience. But in order to know precisely what is best suited to the peculiarities of his soil, climate, and other circumstances, and to know this with the highest attainable certainty, his ultimate reliance must be on his own experience, and that experience he has the means of

10*

enlarging to any extent, by accurate trials, that may be indefinitely multiplied and repeated.

When the best faculties of the owner are thus brought in contact with his soil, a striking and miraculous change is at once visible. Every fertilizer is made richer, every mode of treatment becomes a best method, and all the processes of vegetation are galvanized into new life by the seething battery of his ever-active brain. Any man can follow out a process mechanically when the rules are laid down. But the intelligent, well-read farmer will improve upon the rules, and reach higher results. His success is a matter of philosophical necessity.

Let the cultivator of the soil, then, remember that there is in all this no grand secret nor profound mystery. The triumphs of agriculture are simply the results of patient thought and study. The humblest farmer, whose scanty acres are hidden among the Alleghanies, may communicate to his rock-bound soil the prolific affluence of his thoughtful mind, till every acre shall teem with incredible tons of hay or with unprecedented bushels of corn.

Every increased yield per acre *should* show, and if obtained on sound principles *will* show a diminution in the cost of production. Viewed in this light, a large yield of corn becomes a subject of peculiar interest, and a general and material increase in the acreable product of the country would be equivalent to a reduction of the cost of living for our whole population. At the same time, such an increase in the yield would possess another important significance.

It would be recognized by the hard-working and ill-requited peasantry of foreign lands as a conspicuous inducement to come over to America, and establish themselves on the free and fertile acres of the boundless West.

THE LARGEST YIELD ON RECORD.

The largest amount of corn known to have been produced on a single acre is the yield of Dr. Parker, of South Carolina, which, as mentioned on a former page, was a fraction over two hundred bushels. This quantity of grain does not by any means indicate the highest capability of an acre, but it stands at the head of all known products, and is therefore an event of historical importance.

Yet this result does not seem to have arrested the attention due to its magnitude and its possible consequences; nor has the man who produced it received the full measure of credit to which he is justly entitled. The highest yield ever obtained of a grain that forms the most important crop of the country, if not of the world, is a conspicuous fact in agriculture, that ought to win universal recognition, and confer upon its author a heritage of renown, if not something more substantial.

But the world is often slow to discern, and slower still to appreciate its true benefactors. Successful results gradually developed in the routine of daily life, or in the pursuit of a regular calling, however useful, solid, or lasting, seldom make an immediate or deep impression on society; but individual success, of little or no merit, of no general interest, and no enduring consequence, if suddenly achieved, even without the aid of mental force or moral causes, whether resulting from accident, from impudence, or from crime, raises its author at once to celebrity, and fixes upon him the admiring gaze of the community.

The man who, by a bold and reckless venture in the stock market, gambles successfully and achieves a sudden fortune, is surrounded, as if by magic, with an instant train of admirers. Yesterday he could scarcely claim a friend in the world. To-day his receptions are crowded with the wealth and fashion of the metropolis; he is the centre of observation, and his name is on a thousand lips. He has made a desperate stake, and luck was on his side. Though of the most ordinary capacity, the chances ruled in his favor, and the homage of society rewards his success.

The horse that succeeds in accomplishing his mile a few seconds sooner than any other, wins renown for himself and makes his master a hero. The event excites universal interest, and the press teems with eulogies, that are shared in due proportion between the steed and his owner. The latter, by the fortunate possession of a remarkable animal, is raised to prominence in society, and the suffrage of the community makes him a celebrity whose praise is on every tongue.

The pugilist who, by dint of muscle and power of

endurance, succeeds in vanquishing his antagonist in the ring, punishing him within an inch of his life, and pounding his features into a condition equally frightful and disgusting, is triumphantly escorted from the arena by an applauding multitude, and journals of nearly every rank emulate each other in relating the exploit and lauding the hero, whose fame goes abroad on every wind of heaven, till it spans the whole country.   Such is the reward of the human brute who, by a fortuitous endowment of physical strength, has been able to bruise and batter his unpitied victim to the verge of annihilation.

But here is a man who, quietly and without pretension, has achieved a higher result in the production of food for the human family than any other man has ever reached, who has put on record his two hundred bushels of corn per acre, as a standing protest against the low average yield of the country, thereby making himself the true champion of the cornfield and the genuine hero of productive industry; yet the event has attracted but little attention, and his name is scarcely heard or known beyond his own immediate circle.   Such is the equity of public opinion, and such the civilization of the nineteenth century!

But though Dr. Parker, by his immense and unexampled yield of this grain, has to that extent, and up to the present time, risen above all competition, placing himself, in one sense, at the head of the four million corn-growers of the country, and though his yield, viewed in contrast with the average ratio of production, appears truly prodigious, he has by no means reached the *ultima thule* of possible success, nor demonstrated the yet undeveloped capacity of corn. There is reason to believe, both on theoretical grounds and from actual though limited trials, that the two hundred bushels of Dr. Parker are destined to be materially surpassed, and probably at an early day.

It detracts nothing, however, from the credit of his achievement to know that larger products, on a small scale, have already been obtained. Experimental results, though of limited extent, point clearly to other and still higher yields. While it is true enough that such results may not indicate, with certainty, the product of an acre, yet they are too significant to be lightly regarded. The amount actually obtained from a square rod, however large or incredible it may appear, is prophetic of a similar product for entire fields.

Natural laws can be examined and tested quite as accurately and certainly on a small area of ground as on one of larger extent. The man who obtains forty-four quarts of grain from a square rod renders it probable that either he or others, stimulated by his example, will get two hundred and twenty bushels from an acre. The latter may indeed be more difficult to effect, yet in due time it will be accomplished. And if from an area of half a rod the persevering experimenter succeeds in getting twenty-four quarts of shelled corn, he may fairly claim that he has established, not indeed the fact, but the undoubted possibility of two hundred and forty bushels per acre. He has shown that Nature has erected no inexorable barrier in the way of such a yield, and that therefore how soon it will be reached must depend upon the skill and ingenuity and perseverance of man.

From these and like considerations it is rendered more than probable that some of our thoughtful and progressive cultivators will yet reach a product sufficiently in advance of any hitherto recorded to mark an epoch in corn husbandry.

Whoever the farmer may be that shall first obtain such a yield, if he shall reach it by a method so sound and systematic as to repeat its results, and at the same time reduce the cost of production in a reasonable proportion, he will announce to the world an era of cheaper living, and will deserve to be ranked with the benefactors of mankind. He will increase the money value of every acre of land in the country, and augment the swelling tide of immigration by sending across the Atlantic a new and louder note of invitation that will fall like pleasant music on the ears of toiling millions, kindling in their minds bright visions of future comfort and plenty in the land of Washington and Lincoln.

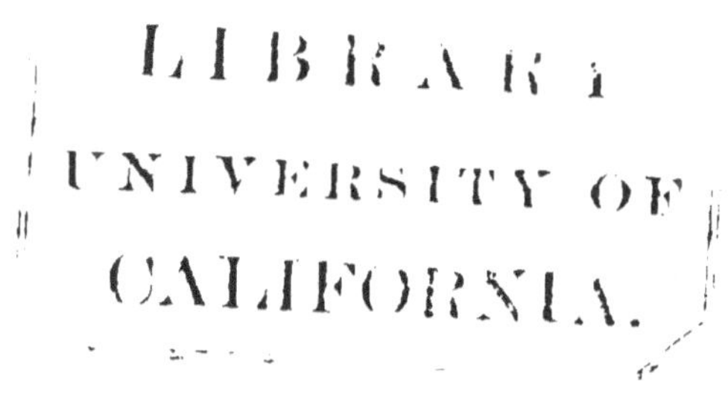

USES OF CORN.

THERE is no plant or vegetable cultivated by man that is capable of being applied to so many purposes of utility as Indian corn. A slight glance at its many and diversified uses is sufficient to show how extensively and intimately it is interwoven with the interests of the human family.

The grain both green and ripe, the stalks and leaves in the successive stages of their growth and maturity, the husks that envelop the ear, and the cob that supports the grain, are all adapted to economical purposes, and fitted, in a variety of ways, to subserve the wants of man.

1. CORN AS AN ARTICLE OF HUMAN FOOD.

The manifold and ingenious preparations of this grain intended for the table, comprise a numerous catalogue of dishes, all differing from each other, yet each possessing its points of merit and its class of admirers.

IN THE GREEN STATE.—In that stage of its growth when the ears and kernels are fully developed, but not yet glazed and hardened, the flavor of corn attains its highest perfection. The ears of the sweet varieties, while yet green, succulent, and juicy, are universally esteemed a luxury, whether boiled or roasted; and the grains, when shaved or grated from the cob, are skilfully converted into a diversity of fritters, cakes, puddings, pies, and other numerous preparations. Some kinds of green corn are thought by many to resemble and rival, when rightly prepared, the flavor of the oyster, and are consequently highly popular with a large class of consumers.

This increasing fondness and demand for the favorite kinds of sweet corn have caused them to be, within the last few years, very extensively preserved by various processes, either of drying, pickling, or canning, which is now so successfully done that the flavor of the green state is retained, and proves highly acceptable on the table at a later period of the year.

Judging from the remarkable and continually increasing demand for some of the leading varieties of green corn during the season of its growth, and the increasing quantities annually put up for winter, it would seem as if the general fondness for it amounted to a passion. The immense supplies poured into our large cities during the summer are almost incredible, and the process of canning it bids fair to grow up into an extensive branch of business. When it is considered that, in addition to the vast amount brought into market, nearly every farmer raises a supply for home consumption, it will be seen that the crop of sweet corn, even by itself considered, forms, both as to amount and value, an important item in the catalogue of farming products.

IN THE RIPE STATE.—But still more variously and extensively in its ripened state, does the grain of this cereal meet the requirements of daily use. In the several forms of hulled corn, popped corn, hominy, samp, Indian meal, corn-starch, and maizena, and in the many simple, healthful, and economical preparations by which these are rendered acceptable to almost every variety of taste, the corn crop of our country is daily contributing, in large and liberal measure, to feed its population.

The public interest in this subject has been from time to time awakened and stimulated by several agricultural journals, and especially by Mr. Judd, in the *American Agriculturist* for January, 1862. The following extract from an article in that number has an historical interest, and is creditable to the enterprise of the proprietor, while it also presents in a favorable light the usefulness of his journal:

"In November last we stated that, taking into account the current prices of corn, wheat, and potatoes, in different parts of the country, West as well as East, and estimating the relative proportion of healthful nutriment furnished by a bushel of each, it seemed evident that a similar amount of nourishment would be obtained from—

> 40 cents expended in purchasing CORN,
> 100 cents expended in purchasing WHEAT,
> 160 cents expended in purchasing POTATOES;

and that, with the present large crop of Indian corn,

and the great foreign demand for wheat, it was especially important to use more corn for food, and save
our wheat to sell. In order to call out information
upon the best methods of *cooking* Indian corn meal,
we proposed, in the December *Agriculturist*, to have
an exhibition of corn bread and corn cake, at our office,
on December 14th. Premiums of ten dollars, five
dollars, and two dollars were offered for the best,
second best, and third best loaves of bread, consist-
ing mainly of corn meal; also an extra premium of
four dollars for the best loaf of cake of any kind in
which corn meal should be the chief ingredient.

"*A Special Cake Premium.*—As the extra premium
of four dollars for corn cake was limited somewhat
by the *cost*, we afterwards decided to add to our published premiums a special premium of two dollars, to
be awarded to the *best corn cake of any kind, without
regard to cost.* The main requisites for the bread
were to be: *cheapness*, fair quality, and adaptability
to *general family use*, eaten cold as well as hot, and
when from one to three days old. Full directions for
making were to accompany each loaf. The entries
reached over *two hundred* (two hundred and nineteen).
Several entries being for duplicated loaves, the entire
number of specimens reached some two hundred and
fifty! As will be seen below, these came from the
distant West, from the Middle States, as far South as
Maryland, and from the North and East. A space of
seventy-four feet of wide table-room was closely filled
with a most imposing display of loaves of all sizes,
from nearly half a bushel down to patty-pan corn-

meal biscuits, and small corn-meal crackers—and not bad crackers either. There were pure corn-meal loaves, and loaves of 'rye and Indian;' loaves one part wheat or rye flour with three parts corn meal, and loaves apprently half meal and half flour, with loaves of every intermediate combination. There were pumpkin loaves, corn-meal dodgers, corn-meal pound cake, corn-meal pone, corn-meal crullers, corn-meal 'nut-cakes,' corn-meal baked puddings, and corn-meal whatnots. There were round loaves, square loaves, high loaves, and flat loaves—in short, loaves of every conceivable form and shape, for of the two hundred and fifty-odd specimens scarcely two were alike in form and mode of making. The sight was one to gladden not only the hungry, but to cheer the heart of every patriot, when he remembers that corn is our native cereal, that it grows everywhere and in abundance, that it is as yet untouched by any disease, that it is healthful and nourishing, and that to-day one with cash can buy, from ready sellers at the West, more bushels of corn at fifteen cents a bushel than there are bushels of wheat now on the continent. The exhibition showed at a glance the great variety of palatable forms in which corn meal can be worked up. Under or by the side of each specimen were placed the directions for making it. The large concourse of visitors, numbering by thousands, were both surprised and gratified, and many went away resolved to henceforth largely increase their family purchases and use of corn meal."

The peculiar fitness of corn for human food, and

its adaptation to the varied wants of the system, have been well stated in the following extract from an article found in several contemporary journals, though we are not certain of its original source:

"During excessive fatigue in low temperature, wheat flour fails to sustain the system. This is owing to a deficiency in the elements necessary to supply animal heat; and the strong desire for oleaginous substances, under these circumstances, has led to the belief that animal food is necessary to the human support. But late scientific experiments have led to better acquaintance with the habits of the North American Indians, and show that vegetable oil answers the same purpose as animal food; that one pound of parched Indian corn, or an equal quantity of corn meal, made into bread, is more than equivalent to two pounds of fat meat.

"Meal from Indian corn contains more than four times as much oleaginous matter as wheat flour; more starch, and is consequently capable of producing more sugar, though less gluten; in other important compounds it contains nearly as much nitrogenous material. The combination of alimentary compounds in Indian corn renders it alone the mixed diet capable of sustaining man under the more extraordinary circumstances. In it there is a natural coalescence of elementary principles which constitute the basis of organic life, that exists in no other vegetable production. In ultimate composition, in nutritious properties, in digestibility, and in its adaptation to the various necessities of animal life in the different climates of the earth, corn meal is capable of supplying more of the absolute want of the adult human system than any other single substance."

In addition to the amount of corn consumed in the various forms and modes of preparing it, both in the green and ripe state as above described, there are also other forms, not perhaps so generally considered, in which it is extensively, though unconsciously, consumed by every class of the people, not only of this, but of other countries. The beef, butter, and cheese, the pork and lard, the poultry and mutton, which make up so large a share of the products of our agriculture, are each composed, in a larger or less degree, of this all-pervading cereal.

When the citizen of a foreign country sits down to a dinner of American beef or pork, the dish before him is the contribution of an American cornfield, representing, perhaps, the golden Sioux of New England, or the stately Gourd-seed of Illinois. The wealthy resident of the metropolis, whose fastidious palate has not, perhaps, been educated up to the latest improvement in corn bread, dilates with complacency over his favorite spare-rib, or tender-loin, without reflecting that the perfection of its flavor is derived from Indian corn.

There are probably few of the consumers of beef, pork, and mutton, who consider the influence exerted by the maize crop on these staples, and fewer still who are fully aware how much higher they would be in price, as well as inferior in quality, if that crop were suddenly annihilated, or even if it were seriously damaged for a single season.

2. CORN AS FOOD FOR DOMESTIC ANIMALS.

Every description of live stock that is usually kept upon the farm may be fed with economy and advantage upon the grain or the stover of maize, or upon both combined, provided these are given with judgment, and not to the exclusion of other feed. For poultry and swine, the grain itself is well adapted. All other kinds of stock will eat with avidity both the stalk and the grain, and will thrive upon them, if they are properly prepared and blended in suitable proportions with other provender.

Corn meal is sometimes fed to cattle without due regard to regularity, and in quantities inconsiderate and unreasonable. The effect of such feeding is not only to injure the animal, but to bring undeserved odium upon the grain. Indian meal is a concentrated feed, and like guano among fertilizers, depends for its highest usefulness and value upon being judiciously blended with the right material, and in the right proportions.

It is a good general rule in feeding, and equally applicable to all kinds of grain, as well as to roots and hay, to confine no class of animals to any one or two articles of food. Variety is conducive to health, and the more carefully the husbandman acts upon this principle, the better his stock will thrive.

FOR POULTRY.—In feeding fowls and most kinds of poultry the rice corn and other small varieties are found to be well adapted and are now generally preferred. Corn-meal and boiled potatoes, mixed together with hot water, are said to be an excellent preparation for feeding to poultry through the winter. To this some add a proportion of oat-meal, and commend the combination very highly, as promotive of health, and increasing the product of eggs. When fowls are to be fattened for the table, they should be shut up for several weeks and fed, four or five times a day, with corn meal and ground barley or oats, mixed together in the proportion of two to one, with warm water and lard. For fattening turkeys, there is no feed like Indian meal, and few if any modes of using corn with more profit. His especial weakness, says a writer in the *Agriculturist*, is Indian corn, and his eye twinkles with delight at the sight of this golden grain. His flesh tells the story of his keeping. For the last six weeks of his life he should be plied with corn as the standard diet. There is no cheating the consumer. A lean bird is not the thing for forty cents a pound. Be honest, give him a plump corn-fed fowl, and sleep with a thriving pocket and a good conscience, though the crib grows lean.

FOR HORSES,—Indian corn, in connection with other feed, is well adapted, and if not given in excessive proportion, is attended with advantage to the animal, as well as profit to the owner. In the livery stables of large towns, and among stage proprietors, the addition of corn meal to the daily feed of the horse is quite generally practised. The proportions usually given are about sixteen to twenty pounds of ground corn and oats daily, with eight or ten pounds of chaffed hay, the ratio of corn to oats being generally about two to one, though this depends very much on the relative prices of these grains. Among farmers this practice may be, and often is, modified with advantage, the chaffed stover of corn being more or less blended with the hay, or substituted for it.

Some men are accustomed to regard oats as the peculiar and essential feed of the horse, without which he can scarcely exist, and with which he needs little besides. It is undoubtedly true, that this grain is well suited, and congenial to the nature of the horse, and no other is perhaps more . so. But this will scarcely justify the practice of making oats his exclusive feed, nor of limiting his diet to oats and hay. According to principles of physiology, as well as on evidence derived from experience, the horse, like every other animal, requires variety in his food, and cannot without it maintain a condition of perfect health and vigor.

VALUE OF CORN FOR CATTLE.—In the management and feeding of neat cattle, there are several classes of them to be considered; namely, young stock, milch cows, working cattle, and beeves. For each of these, Indian corn is found useful, and if the object is to produce the highest degree of thrift in the animal at the least expense to the owner, and to support the largest amount of stock on a given extent of ground, then Indian corn becomes not only useful, but indispensable. Every part of the plant, including the leaf, stalk, husk, and cob, as well as the grain, may be turned to an advantageous account.

For young stock, and for cows, when milk rather than butter is the object, the stover alone, if well cured, finely chaffed, and soaked a few hours before feeding, is sufficient to keep them in good condition; though for the purpose of variety, it is usual and profitable to connect with this a proportion of cut hay, or pulped roots, or both. But for working cattle, for cows when butter or cheese is the object, and for beef-cattle at all times, the grain is essential to the best results, and should be combined with other kinds of feed in larger or less proportions, according to circumstances.

For this purpose, there is perhaps no better preparation of corn-fodder than that already described on a previous page, namely, the addition of corn and cob meal to the stover finely chaffed. This combination includes the entire product of the corn, and when thoroughly scalded or steamed before using, whether given for the purpose of butter or beef, or for the general improvement and vigor of the animal, is found to be exceedingly well adapted to the intended object.

It was shown, in a former chapter, that when the farmer raises one hundred bushels of corn per acre, the total product of the crop, in the form of this fodder, is fifteen thousand pounds, and is equal, in nutritive value, to twenty thousand pounds of hay.

Now it has been found in practice, that cattle require, for their daily food, from two to three per cent. of their weight in hay, or its equivalent. According 'to Prof. Johnston:

An ox at rest requires 2 per cent. of its live weight.
 " " at work   "   2½ "   "   " " "   "
A cow in milk   "   3 "   "   " " "   "

If then, we take the average weight of cows at seven hundred pounds, it will appear that the above product of one acre of corn would be more than sufficient to winter six of them, assuming the average winter for the United States to be one hundred and fifty days. Or, taking cattle of the various classes, at an average consumption of food equal to two and one-half per cent. of their weight, then the above product of one acre of corn would support seven during the winter, and leave a balance of the crop on hand.

VALUE OF CORN FOR SWINE.— In feeding hogs, the stover of corn is not, to any considerable extent, available, and the grain is usually given in larger proportion, and more exclusively than to cattle. But here, as in other cases, the principle of variety is not to be neglected. In diversifying the feed of this animal, there need be no difficulty. His omnivorous propensities are so strongly developed as to embrace nearly every kind of food that comes within his reach. Whatever is eaten by other domestic animals is seldom rejected by the hog, while many substances refused by them are eagerly appropriated by his indiscriminate voracity.

He is, therefore, easily kept, and with prudent

management may be made very profitable. This is especially the case where pigs are kept in small numbers. When the proportion of them to other stock is rightly balanced, so as to make them a convenient appendage to the barnyard, to the kitchen, and to the dairy, the cost of maintaining them is so trifling as scarcely to be felt.

When they are increased beyond this proportion, though the cost of keeping the additional number is somewhat greater, yet with good management the comparative expense may be made very moderate. The incidental and economical sources of food for swine on a well-managed farm are so many and various that very little positive expense is incurred, except for the grain that is superadded to the other feed in the process of fattening.

But here is where many farmers make a serious mistake. They postpone the use of grain until a late period, and then commence feeding it in excessive quantities, that are often suddenly increased with but little regularity and little or no system.

With such treatment, neither corn nor any other grain or feed can exert its proper and legitimate effect. Experience has proved that the most certain mode of feeding hogs, with profit, is to commence the use of grain or meal with the young pig. It matters not how young the pig may be, provided the meal is given with due care and judgment, in small quantities, well scalded or cooked, and fed in connection with milk and other waste from the dairy and kitchen.

If this practice is continued with a *very gradual*

increase in the amount of Indian meal until the time arrives for full feeding, the farmer will find his hog so far advanced in size and flesh that a much smaller quantity of grain will be required to finish off with, than would be needed by the other method.

He will thus have a healthier animal, better pork, and more of it, with a *less total consumption of corn*, than he could obtain by the mistaken system which vainly attempts to compensate for the neglect of the first six months by excessive feeding during the last two.

This principle is equally true and sound in reference to other animals, and a similar economy of grain and increase of flesh will be found to result from it.

CORN FOR SHEEP.—In the feeding of sheep both the stalk and the grain of maize may be used with advantage, especially when blended or alternated with other kinds of feed. There is, perhaps, no animal that thrives better on a variety of food, and none that needs more careful attention, both in the feeding and in the general management.

It is a very usual practice with sheep farmers to mix Indian corn with oats or barley, giving to each animal a pint per day, in addition to other feed. The latter grains are doubtless good, and rank probably next to corn in value. But when the object is to fatten the animal there is no feed equal to Indian meal and oil meal, given in equal quantities, and not less than one pound per head daily.

According to Mr. John Johnston, of Western New York, a most excellent authority, there is no animal that will take on fat so rapidly as sheep, if they are in good condition and rightly fed. He estimates the relative values of oil meal and corn in the ratio of fifty pounds of the former to sixty pounds of the latter, in feeding sheep, and probably the same comparative value would hold true for other animals.

It has been found by wool-growers that the tendency of feeding corn is to increase the weight of the fleece. One writer reports, as the result of some experiments on this subject, an average increase of half a pound per fleece, produced by feeding corn during the winter, before shearing. Some others have found a greater increase than this.

Sheep are very fond of corn-fodder when it is perfectly sound, and experience has proved it to be economical, and well adapted to their wants, and all the more so if finely chaffed. The amount of food consumed by sheep, as compared with that of cattle, has been computed at about one-eighth. That is to say, the food required by one ox would be sufficient, on an average, for eight sheep. Therefore, according to the estimate before made for cattle, it will be seen that the total product of one acre of corn, including grain, cob, and stalk, on a yield of one hundred bushels, would be sufficient to winter more than fifty sheep.

It is not, however, intended by this statement to recommend the exclusive use of the above provender; nor to prescribe for every case one invariable proportion of the different parts in combining them. Though for some purposes the combination of the grain, cob, and stalk, in the proportion of their yield as above given, is no doubt the best form in which they can be used, yet, in other cases, a larger or less relative quantity of the grain would be found expedient. These are points that the farmer can best determine for himself by comparative trials. The estimate is made for the purpose of illustrating the economical advantages of corn, and the feeding power of one acre of it.

COST OF BEEF MADE FROM CORN.

THE neat cattle in the United States, in 1860, including all kinds, amounted to over twenty-five millions. In 1865 the beef consumed in New York city alone, added to the quantity exported from it, made a total of over one hundred and fifty million pounds. What the entire consumption and export of beef for the whole country amounts to, we have no means of determining; but from the amount required by a single city, it is easy to see that the aggregate demand is immense, and that, to supply this demand, the production of beef by American farmers has grown into a business of vast proportions.

As Indian corn is a large element in the making of beef, the best method of feeding it becomes, of course, an important question, and interests alike the producer and consumer. It interests the former by determining, to a certain extent, the amount of profit on his corn and other provender, and the latter because it involves the cost, and therefore affects the price, of an article of daily consumption.

The experience of farmers in regard to the profit

11*

of making beef is widely various, but, on the whole, unfavorable. One man finds the business lucrative, while another sinks money in it. The difference arises in part, no doubt, from the locality, the breed of the animal, and other circumstances; but it also depends very much on the method of feeding, and on the man.

If a few invariably succeed, or even generally succeed, although a larger number may fail, it proves that there is a right method that brings success, and that consequently success ought to be the rule, and failure the exception. No man who proceeds blindly in this business can reasonably expect to make it profitable. It is as true here as in every other branch of husbandry, that intelligence is essential to prosperity.

In order to convert corn, or any other feed, into beef to the best advantage, it is important to know, as nearly as possible, how many pounds of the former it requires, on an average, to make a pound of the latter. This does not appear to have been, as yet, very precisely determined, in regard to corn; there are, however, some data from which a tolerably accurate conclusion may be derived.

There is also another principle, now beginning to be understood among farmers, that should here be kept steadily in view. It is found that a certain amount of food is consumed by every animal before the process of fattening commences. When a steer is brought up to the point where this process begins, it requires a definite quantity of provender to keep him in that condition. If fed beyond that point, the *excess* of food contributes to the formation of fat. Thus in regard to beef, as we before found in the case of corn, the profit lies in the last additions made to the cost of production.*

Here, then, arises the twofold question, What is the amount of provender that will keep an animal stationary ? and what amount of corn or other feed, in addition to this, is required for each pound of fat that is formed ?

Now we have already seen that neat cattle consume, on an average, two and a half per cent. daily of their weight in hay, or its equivalent. If they receive less than this, they fall away ; if more than this, they increase. If, then, a steer weighing seven hundred pounds is fed one hundred and twenty pounds of hay, or chaffed stalks, per week, or any other food *equivalent* to these, he will hold his condition. If, in addition to this, he receives fifty-six pounds of corn per week, he will increase in weight. In order to know definitely what the gain would be in this case, let us endeavor to determine the effective value of corn in the production of beef.

Mr. G. H. Chase, of Cayuga County, N. Y., found by experiment, as reported in the *Country Gentleman*, that twenty-eight quarts of ground barley per week gave an average increase of eighteen pounds of flesh ; but ground barley contains less than one-fourth the percentage of fatty matter that belongs to Indian

* This principle is equally true in all cases of feeding, whether the object is beef, butter, cheese, pork, or mutton.

corn, and the latter has been proved by trial to be more fattening than any other grain. It appears, from the experiment of Mr. Chase, that less than three pounds of barley gave one pound of beef. This, however, is probably better than an average result.

In the Journal of the Bath and West of England Agricultural Society a table is given, in which six pounds of barley are stated to be equal to the production of one pound of beef.

In some experiments on pig-feeding, by Mr. Lawes, of England, the comparative fattening effects of barley and corn were found to be very nearly in the ratio of six to five; making five pounds of corn equal to six pounds of barley. Therefore, according to the table above referred to, five pounds of corn would be equal in feeding effect to one pound of beef.

In the *Rural Annual* for 1865, the editor, commenting on some experiments of Lawes and Gilbert, comes to the conclusion that "a bullock weighing eight hundred pounds would consume forty-three pounds of corn and ninety pounds of hay per week, and increase eight pounds."

It is evident that this rate of feeding is entirely too low for the weight of the animal. It shows a fair result for the corn, but too small a gain of flesh to give the highest profit. It would take all of the above hay, and about half of the corn, to keep the ox stationary through the week, and the balance of the corn, say twenty-four pounds, would produce the increased weight.

But the writer afterwards varies this statement,

and supposes one bushel of corn and one hundred pounds of hay to produce ten pounds of beef in a week. But still the rate of feeding is too low for the best result. In this case the ox would require the entire hay, and about twenty pounds of the corn, in order to hold his condition; leaving forty pounds of the corn to account for the increased weight. In one of these instances the effect of the feeding shows that three pounds of corn produce a pound of beef, and in the other four pounds of corn give the same result. In both cases, if more corn were given, it would increase, not only the gain of flesh, but the rate of profit on the animal.

According to the principle stated by Mr. Lawes, and established by his experiments, it seems evident that, with a good breed of cattle, from three to four pounds of corn, in addition to the above proportion of other provender, will give a pound of beef.

There are those who consider the effective value of corn even higher than this, while others place it quite as much below these figures. On the whole, we think it may safely be assumed, that, after the animal has received the amount of food necessary to sustain it, every four pounds of corn in addition will give one pound of beef, provided the meal is properly fed, by being well mixed with the other provender, and thoroughly soaked or steamed.

Now, taking the case of the steer weighing seven hundred pounds, let us see what the beef would cost per pound by this estimate. The amount of feed per week was assumed to be one hundred and twenty pounds of chaffed stalks, to keep up his condition, and fifty-six pounds of corn to fatten him. We now find that fifty-six pounds of corn would give an increase of weight to the animal equal to fourteen pounds. If we suppose the farmer to charge his corn at one dollar per bushel, and his stalks at six dollars per ton, the account would stand thus :

56 lbs. of corn............................	$1 00
120 " " stalks.........................	32
	$1 32
Deduct value of manure.............	60*
	72

The farmer here gets fourteen pounds of beef at a cost of seventy-two cents, which is equal to five and one-seventh cents per pound, while the profit on his corn and stover is, or ought to be, at the above prices, over one hundred per cent.

But to illustrate the principle above referred to, and to show the effect of higher feeding upon the rate of profit, if we suppose the quantity of food increased in the above instance, in the right proportions, it will be found that every additional pound of beef, made by such increasing in the amount of feed, will cost but four cents; and if the corn and stover were charged at the cost of production, instead of at the figures above given, then the cost of the beef thus added would be about two cents per pound.

* Some farmers consider the manure of a well-fed steer equivalent to $1.00 per week.

This case of feeding, which is given as an illustration merely, would not be strictly followed in practice, as a greater variety of food would be better for the animal, and would not materially alter the result. Pulped roots may always be used with advantage in connection with corn-meal and stalks, if the proportion is properly regulated.

It will be seen that if the farmer, in this instance, sells his beef at cost, he gets one dollar per bushel for his corn, and six dollars per ton for his stalks, out of which, however, is to be deducted the cost of grinding the grain and chaffing the stalks.

But the price of beef, in the New York market, has not been as low as five cents, on a yearly average, for a long time. The price for the last year (1865) averaged about eleven cents, and for the last six years about seven cents per pound, for the live weight.

If, then, he sells his beef at the average price of the last six years, he realizes for his corn one dollar and fifteen cents per bushel, and for his stalks eight dollars per ton; while if he gets for his beef the average price of the last year, it pays him one dollar and sixty-one cents per bushel for his corn, and ten dollar per ton for his stalks.

The following table indicates the price realized by the farmer for his corn, for different prices of beef, and also for different amounts of corn required in feeding, to produce a pound of beef. Fractions are here omitted, as the results in whole numbers are sufficiently accurate for general purposes :

RATIO OF CORN TO BEEF.	PRICE OF BEEF.	PRICE REALIZED FOR CORN.	
		Grain per bushel.	Stalks per ton.
Five lbs. of corn producing one lb. of beef.	5 cts.	$0 84	$6 00
	7 "	1 01	7 00
	9 "	1 18	8 00
	11 "	1 40	8 00
Four lbs. producing one lb.	5 cts.	$0 93	$7 00
	7 "	1 15	8 00
	9 "	1 43	8 00
	11 "	1 61	10 00

It is here apparent that if it takes five pounds of corn, in addition to the other feed, to produce a pound of beef, the latter, even at five cents a pound, pays eighty-four cents per bushel for the corn, and six dollars per ton for the stalks. Now, if the farmer's corn costs him thirty cents per bushel to produce it, which is about the average cost of production for the whole country, then it leaves him a margin of fifty-four cents per bushel, out of which he can pay for grinding the grain and chaffing the stalks, and a profit will still remain.

But if he succeeds in raising his corn at a cost of twenty-five cents per bushel, and converting it into beef at the rate of four pounds for one, both of which are entirely possible, then at the average market price of beef for the last six years, he makes a profit on his grain of ninety cents per bushel, while the margin of profit on the stalks will pay for grinding the former and chaffing the latter.

But there is another contingency in regard to beef which the farmer may avail himself of with decided advantage. The price of it varies with the condition of the animal. This is an important consideration, and too often overlooked. A very fat steer will bring a higher price per pound than a lean one, or than one even moderately fat. The excess of weight produced by continued high feeding is supposed to impart an extra value to the whole animal. The accession of fat produced by the last ten or twenty bushels of corn not only brings its own higher price, but, at the same time, raises the price of the entire carcass.

This final increase in the fleshiness of the animal seems to convert the beef from an article of necessity into an article of luxury, and carries with it a corresponding change in the market value. Whether or not there is any sufficient reason for this distinction, is not for the farmer to inquire. It is not his province to determine what *ought to be*, but to shape his business according to what *is*. The feeder, therefore, who judiciously takes advantage of this well-known fact, may generally realize from two to three cents a pound more for his beef than the figures in the table.

On the whole, then, it may fairly be assumed that the farmer who makes *good fat beef* may reasonably calculate on getting eight cents a pound for it, on a yearly average. In that case, if he converts his corn into beef even at the rate of five pounds for one, allowing his corn to cost thirty cents per bushel, and his stalks three dollars per ton, it will bring the cost of his beef to about four cents per pound, even without taking the manure into account, and the profit on his corn will be forty-three cents per bushel.

But if he makes four pounds of corn (in addition to the other feed) produce a pound of beef, and counts his manure at its true value, then he realizes a profit on his corn of ninety-five cents per bushel, and on the stover of five dollars and fifty cents per ton; which is nearly the same thing as five dollars a ton for the stalks and one dollar a bushel for the corn. These figures, for an average profit, ought to be satisfactory. Some farmers have done better; and every man who finds his profit falling much below this, has reason to suspect that there is something wrong either in his method of raising corn, or in his method of feeding it.

There is probably no part of the farmer's occupation that requires more careful and constant attention than the feeding of his stock, and none that depends so much for success upon the exercise of intelligence, good sense, and sound judgment.

"Cattle feeding," as the *Springfield Republican* very justly remarks, "is a science of trade, to be studied and learned like any other. Qualities and quantities are not the only things requisite in the care of domestic animals. Regularity, cleanliness, comfort, and quiet repose are elements of thrift, not to be lightly considered. In the application of these is shown the skill of the herdsman. One man will make the same amount of feed go further and accomplish more than another. A great deal depends on knowing how. A herdsman does not become full fledged instantaneously. Among the first steps in progress are the consciousness of ignorance, and the desire to learn."

COST OF PORK MADE FROM CORN.

THE grain that is usually and almost exclusively employed in this country for fattening pigs, is Indian corn. It is found to be more efficient and economical than any other, and imparts to the pork an unrivalled solidity and flavor. Other grains in smaller quantities are sometimes mixed with this, and if the proportion is not too large may be employed to advantage.

Corn that is fed to swine should invariably be ground, and the meal steamed or boiled before feeding. Its nutritive effect and fattening power are surprisingly increased by this treatment, and the practice of the most successful feeders has proved its utility so clearly as to place it beyond any doubt.

It is found that corn is more effective when fed to hogs than in the case of neat cattle, and produces a larger amount of pork than of beef for each bushel consumed. Successful farmers have not unfrequently obtained a pound of pork by feeding from two to three pounds of corn. The gain of flesh per day with good feeding will reach from one to three pounds, and has been known to reach three and a half pounds. It is almost incredible how cheaply pork may be produced with a *good breed of hogs*, if well fed and well managed.

Mr. J. Sibley, of Wayne County, N. Y., has reported to the *Country Gentleman*, that four hundred and twelve pounds of pork, made mostly from corn, cost him twelve dollars and ninety-three cents, which is a trifle over three cents per pound. If the value of the manure had been reckoned in this estimate, as it ought to be, the cost of the pork would have been between two and three cents per pound.

Nathan G. Morgan, of Union Springs, N. Y., as stated in *Tucker's Annual Register*, considers the value of corn doubled by grinding the grain and scalding the meal, and finds that, at five cents per pound for pork, he gets one dollar per bushel for his corn.

William Van Loom, in a communication to the *Prairie Farmer*, says that he has practised feeding boiled corn, and is satisfied that one bushel thus prepared is equal to two bushels fed raw. In one experiment he found that three pounds of cooked corn gave one pound of pork.

Gates Henry, of Schuylkill County, Pa., has stated in a prize article to the *Agriculturist*, that by feeding his hogs fifteen to twenty bushels of corn each, he has usually made the weight from four hundred to five hundred pounds. He does not state that the whole of this weight was produced by the corn exclusively, yet it is evident that the corn was converted into pork at a handsome profit, bringing the cost of the latter to a low figure.

"A very successful manager," says the editor of the *Country Gentleman*, "with whose treatment we are well acquainted, pours six parts of hot water on one part of ground Indian meal, and then allows it to stand twelve to eighteen hours, until the whole is swollen to a thick mass, when it is given to the animals. He finds boiling water better than cold for this purpose, but the mixture undergoes little or no fermentation. So successful is his management, that in connection with the selection of good breeds, and regular feeding and cleanliness, he usually obtains one pound of pork from feeding three pounds of corn."

Mr. J. W. Zigler, of Indiana, according to a statement made by him in the *Western Rural*, fed fifteen hogs with corn for forty-two days, during which time the average gain per hog was nearly three pounds per day, and the pork was at the rate of one pound for every three pounds of corn. The pork was sold in Chicago at ten and a half cents per pound, giving him a net profit of one hundred and forty dollars.

Mr. Baldwin, an English breeder of some note, has used Indian corn, barley meal, and ground peas in fattening hogs, but gives the preference to the corn. He finds that *two pounds of it will produce a pound of pork.* This result is higher than usual, and is probably in part due to the breed of the animal.

Though most of the above figures are better than the average experience of feeders, they might generally be equalled, and some of them surpassed by a majority of farmers, if more careful attention were given to the subject.

It will be found, if the value of the manure is taken into the account, that when three pounds of corn produce one pound of pork, the latter, at six cents a pound, pays one dollar and twenty-eight cents per bushel for the corn. As the average price of pork, for the last six years, was over six cents per pound for the live weight, there seems to be no reason to believe that it will be below that figure for some years to come.

For 1865 the yearly average for pork was over twelve cents per pound. At this price, the farmer who makes three pounds of corn equivalent to one of pork, gets two dollars and forty cents per bushel for his corn, which is certainly a rate of profit that in most kinds of business would be deemed very satisfactory.

The following table gives the prices realized for corn at several different prices for pork, and for different ratios of corn to pork in feeding. The manure is rated at six dollars and fifty cents for each ton of feed consumed, which is about the usual estimate, though less than its real value to the farmer who rightly uses it:

RATIO OF CORN TO PORK.	Price of pork.	Price realized for corn per bushel.
Four pounds producing one pound.	5 cts.	$0 86
	6 "	1 00
	7 "	1 14
	8 "	1 28
Three pounds producing one pound.	5 cts.	$1 09
	6 "	1 28
	7 "	1 47
	8 "	1 65
Two pounds producing one pound.	5 cts.	$1 56
	6 "	1 84
	7 "	2 12
	8 "	2 40

As there is, at the present time, an unusual scarcity of hogs in the United States, there is every reason to believe that the range of prices for pork will rule higher, for some time to come, than the average of the last six years.

The farmer, therefore, who converts three pounds of corn into one pound of pork, allowing the corn to stand him in thirty cents per bushel, which is more than it ought to, will bring the cost of his pork at less than two cents per pound, with a prospect of realizing not less than seven cents, which will make the profit on his corn one dollar per bushel, without counting the manure.

The amount of pork required to meet the demand for consumption- and export, may be partly judged from the fact, that the total receipt of hogs in New York city for the last year was about six hundred thousand, and the amount exported from the same city was nearly one hundred and twenty thousand barrels. As the demand is likely to increase more rapidly than the supply, farmers will probably find it their interest to augment their stock of hogs, and turn them to the best account by feeding them up to a heavier weight than usual, before sending them to market.

12

COST OF MUTTON MADE FROM CORN.

According to the latest opinions and experience of sheep-farmers and others, it seems to be generally concluded, that corn is quite as well adapted for making mutton, as for beef or pork. In the absence of definite experiments, it is not easy to determine the precise value of this grain in the production of mutton; but in a comparative view, and reasoning from analogy, we have ground for believing that, under favorable conditions, three pounds of corn will produce a pound of mutton.

According to Mr. Sanford Howard, " it has been proved that a given quantity of meat can be produced from the sheep at as little, and in some cases less expense, than from any other animal; and so far as can be ascertained, the meat is fully equal in nutritive properties. Here, then, we have from the sheep at least an equal amount of meat, as compared with any other animal, for the food consumed, while we obtain the fleece as clear gain."

It is stated by the editor of the *Agriculturist*, that mutton is more economically made, and more advantageously used up than pork or beef. He also further adds, that " more grain is required to make a pound of pork than a pound of mutton," and that the latter " is more nutritious, and will consequently give a laborer more strength than pork." These statements are, no doubt, entirely true, and if true, are very important, and ought to be more generally understood and acted upon.

At the present time, when hogs are more than ordinarily scarce, it is certain that mutton can be made, with prudent management, at a handsome profit, and the occasion is favorable for inaugurating a more general, if not universal, consumption of this healthful and nutritious food.

It is at least a reasonable presumption, that an animal carrying with it, like the sheep, a twofold source of profit, in its mutton and its wool, ought to be turned by the farmer to a very lucrative account, provided his attention is duly divided between the two objects, and not entirely monopolized by either. Indeed, it may be taken for granted, that whenever the sheep, with its double value of fleece and flesh, fails to prove highly remunerative, there is mismanagement somewhere, and it is highly probable that some part of the fault lies in the feeding.

But in addition to the value of the fleece, another advantage in making mutton is found in the superior quality of the manure. Mr. Johnston, of Geneva, who has been very successful in feeding sheep for the mutton, considers this source of profit a very important feature of the business. There is probably no land so poor, no soil so hopeless, that it may not be restored under a system of sheep-husbandry. There seems to be a natural antagonism between a poor soil and a flock of sheep. Wherever the latter goes the former disappears. Sterility of land flees from the presence of these useful animals, and the invasion of an unfertile region by the shepherd and his flock is the unfailing harbinger of green meadows and prolific fields of grain.

Taking into account, then, the value of the manure, and the value of the fleece, it is more than probable that whenever the cost of producing mutton is fully and fairly tested, by accurate experiments in feeding, it will be found a cheaper article of food than is at present suspected. It will also probably be found that it can be made at a less expense, and of better quality, from the grain and stover of corn (with a due proportion of other feed), than in any other way.

The weight and quality of the fleece varies, of course, with the breed. On a comparison of those breeds that are preferred for their flesh, the average value of the clip would doubtless cover half the expense of feeding, and still leave a fair profit on the wool. In the opinion of many, the fleece would give a better result than this. The value of the manure is probably equal to one-fourth of the expense of feeding, and the remaining fourth represents the cost of the mutton.

Probably the value here assumed for the manure will, by some, be considered too high, and that for the fleece too low. If so, one would offset the other, and the result would still be the same.

Then, if we assume that in feeding sheep four pounds of corn will produce, on an average, one pound of flesh, though it is nearly certain the result would be better than this, we shall have one pound of corn as the cost of a pound of mutton.

Supposing the corn to cost the farmer thirty cents a bushel, this would bring the cost of the mutton at half a cent a pound, and if we add for attendance, etc., as much more, the entire cost would be one cent, which, after allowing a liberal profit to the farmer, would still leave this meat accessible to the million, at a price that would render it the most economical, as it is the most healthful, description of animal food.

In computing four pounds of corn, in the above estimate, as equal to one pound of flesh, it is not, of course, designed to make corn the exclusive feed. The principle intended to be illustrated may, perhaps, be more clearly stated as follows: In the use of any variety of healthy food, judiciously blended, and *comprising a due proportion of corn*, an amount of it equal in nutritive value to four pounds of corn will produce a pound of flesh.

COST OF BUTTER AND CHEESE MADE FROM CORN.

It would be natural to conclude, from the essential nature and quality of Indian corn, that it must be well adapted to the production of butter. This conclusion is confirmed by chemical investigation, and is further ratified by the results of experience.

The first and most obvious effect of corn meal is to improve the *quality* of the milk, and make it richer, by imparting to it a larger proportion of the constituents of butter and cheese. How far it affects the quantity of milk, as compared with some other kinds of feed, has not been very definitely determined. But for improving the flavor and increasing the amount of cheese and butter, it is found to be well adapted, and is thought by many to excel most kinds of feed.

It is also found that the stover of corn has the same general tendency as the grain, though in a less degree. Its most favorable effect and highest value are only realized when the object is especially to produce a copious flow of milk. For this purpose, the succulent stalk of Indian corn, whether fed green in summer and fall, or well cured, chaffed, and steamed, in winter, is probably not surpassed, if equalled, by any provender in use.

Thus the combined result produced by the different parts of corn, one having a special influence on the milk, and the other a similar effect on the butter and cheese, seems to indicate the peculiar fitness and value of this cereal for the purposes of the dairy.

What amount of this feed would be required for a given quantity of butter, has not yet been very accurately determined. Some estimates have been made rating the effective value of corn at from five to eight pounds for producing one pound of butter. In some experiments that have come to the knowledge of the author, the result was less than five pounds of corn for one of butter.

Comparative estimates have also been made as to the relative amounts of beef and milk resulting from a given quantity of feed. Sir John Sinclair, as cited by Professor Johnston, has stated that the same provender which gives one hundred and twelve pounds of beef will yield three thousand six hundred pounds of milk. But this is undoubtedly erroneous; the disproportion in favor of milk being greater than experience warrants us in crediting.

The estimate of Riedesel, a Continental writer, is rather more reasonable, but still not accurate. According to the latter authority, the hay that gives one hundred pounds of beef will give one thousand pounds of milk. Allowing twenty pounds of the latter for one of butter, which is about the general average, this would give fifty pounds of butter from the same feed that produces one hundred pounds of beef. This estimate, though it comes nearer than the previous one, errs in the opposite direction, and the truth undoubtedly lies between them.

Others have computed the ratio of butter to beef, on equal quantities of feed, as eighty to one hundred, which is evidently more reasonable than either of the others, and seems to be very nearly correct. Comparing this with the proportion of beef to corn, as given on a former page, it will be found that, for a pound of butter it would require five pounds of corn, over and above the stover, or other feed given to sustain the cow. Then, by the same calculation that gave fourteen pounds of beef for seventy-two cents, we shall have eleven and one-fifth pounds of butter for the same sum, which is about six and one-half cents per pound.

In this calculation the farmer has charged his corn at one dollar per bushel, and his stalks at six dollars per ton. If these were charged at the expense* of producing them, the effect would be to bring the cost of the butter to about four cents per pound, without taking the manure into the account. If the expense of grinding the corn and chaffing the stalks were added to this, and also the expense for labor in making the butter, the cost of the latter would still not probably exceed six or seven cents per pound.

In dairies devoted to cheese, the total product of

* Calling the expense thirty cents per bushel for the grain, and three dollars per ton for the stalks.

this article per cow is much larger than that of butter, and the relative value proportionably less. It is found that on the same amount of feed, a cow will produce from two to three times more cheese than butter. This ratio is not uniform nor constant, but varies with the breed of the cow, etc. On a general average, it is estimated by many farmers, that a cow will give two and a half pounds of cheese for one pound of butter. Some others make the proportion about two to one. If we assume the latter to be the true proportion, it will bring the cost of cheese to three and one-fourth cents per pound, when the farmer charges his corn at one dollar per bushel, and the stalks at six dollars per ton, or to two cents per pound, when the corn and stover are charged at the cost of production. After a fair allowance for the expense of labor in preparing the feed, making the cheese, etc., it would probably be found that the cost of the latter would be about four or five cents per lb.

The above estimates for butter and cheese are based on the methods of making them usually practised by farmers. But recent improvements have been introduced, and plans adopted, that have a tendency to modify and reduce the cost of production in the case of these articles. They are now extensively made by associations that have proved remarkably successful in producing both cheese and butter, especially the former, at a great advantage and with diminished expense.

The foreign demand also that has recently sprung up for cheese made from skimmed milk cannot fail to have the effect of increasing the profit on butter, by enabling the farmer, after the cream is taken from the milk, to turn the latter to a more lucrative account than formerly.

Yet these circumstances can have no material effect upon the principle on which the above estimates are based. Whatever changes may be made either now or hereafter in the plan of making butter and cheese, yet the modes of feeding, the varieties of food, and the proportions of them, remain the same. The principles of feeding that we have endeavored to illustrate, as well as the relative value of corn, and the advantage of using it in due proportion, if found correct in one case, will prove equally so in the other.

HOW TO MAKE FEEDING PROFITABLE.

It is generally understood, and appears from the preceding investigation, that when the feeder sells his beef and mutton or the dairyman his butter and cheese, whatever the price they bring, and whatever the margin over the cost, there are two classes of expenses to be deducted, and two separate profits to be secured. When the farmer realizes on the sale of his butter, beef, and other products, a net result that gives him a fair profit on these articles over the cost and care of feeding, and a similar margin on the corn and other feed by which they were produced, he closes up the business of the year with a satisfactory balance on the right side of the account, and is entitled to consider himself a successful man.

If, on the other hand, he discovers, on the sale of those products, that the expense of feeding and the cost of cultivation have not been reimbursed, and that after a year of toil there are no net gains to be counted, he may justly suspect, that in some one or several of the processes and operations of his farm there has been either culpable neglect, or needless and inexcusable want of information, or very possibly both.

He may vainly attempt to divide the blame between an incorrigible soil on the one hand that refuses to reward a slovenly mode of culture, and an obstinate class of animals on the other, that do not choose to fatten upon neglect: but if he will reflect upon the nature of his business, and consider how many separate and distinct operations there are upon which the profit of his butter and beef mainly depend, he will find his want of success easily explained.. He will discover that, in all the different processes from which pork, mutton, and beef are the *final* result, no one of them can be overlooked or disregarded, without some diminution of his ultimate profits.

This important reflection, though seldom duly weighed, deserves the serious consideration of every man who cultivates the soil. Between the planting of the corn and the slaughtering of the ox there are more than a score of separate operations, each one of which produces an effect on the cost of the beef.

If the farmer plants his corn a little too deep, or too late in the season, or too close together, or too far apart; if he applies the wrong kind of manure, or the wrong quantity, or at the wrong time, or fails to apply any; if his ground is imperfectly ploughed, or ploughed at the wrong time; if he handles the horse-hoe carelessly or too seldom; if his corn is cut out of season or defectively cured; if it is fed to his animals in an unsuitable condition, neither ground, cut, nor steamed; if they are fed too seldom, or too much at one time and too little at another; if the feed is deficient in variety, or combined in the wrong proportions;—each one of these separate contingencies, as well as many others not mentioned, exerts its own peculiar influence, small in some cases but great in others, upon the cost of beef, pork, and all similar products, and each one of these helps to determine the question whether the final result will be a profit or a loss.

Thus it appears that the cost and the profit of these products have already begun to accrue when, in early spring, the farmer strikes the first furrow in his cornfield, and the plough in his hand becomes a mathematical instrument that helps to solve a question of figures. It may, in fact, be said with truth, that still earlier than the spring this question of cost has begun to be solved. When in the previous fall the cultivator goes into the field to select his seed-corn for the following crop, even then he settles, in that brief interval of time, one of the important contingencies on which his future profits are suspended.

Considering, then, how many distinct operations the farmer goes through, before reaching his final results, and how certainly these results are affected by each operation and by his manner of performing it, it is scarcely surprising that experience differs so widely in regard to the profit of feeding. It would seem that in farming, as in every other business, success depends, after all, more upon the man than on any other cause. Some men are constantly seeking information and accumulating knowledge, while others prefer to cleave to their ignorance. One man contrives to do every thing nearly right, while another is equally infallible in doing every thing nearly or quite wrong.

It is easy, then, to perceive that, if in the production of butter or beef there are twenty or thirty different processes to be gone through, and one man adopts the best method in each, while another performs each imperfectly or not at all, their experience in the end will be entirely opposite ; one making a certain profit and the other incurring inevitable loss. It is quite possible that each one of these various processes might make a difference, on an average, of nearly one cent a pound in the cost of beef or mutton, and of several cents per bushel in the cost of corn.

If, then, every farmer who embarks in feeding stock for market, or in making butter or cheese, would adopt the obvious course suggested by these reflections, giving careful attention to each particular process all the way through, and making sure that each one is rightly performed and at the proper time, he would find that feeding can be made a profitable business, and that by using all his faculties, mental as well as physical, his success would be morally certain.

MISCELLANEOUS USES OF CORN.

THOUGH the principal value of maize is due to its nutritive property, and its highest importance lies in the amount and quality of the food it supplies, there are yet other and various economical purposes for which the several parts of it have been found to be well adapted.

PAPER AND CLOTH.—Many attempts have been made, with various success, to use the fibre of corn in the manufacture of paper. This fibre is contained in the husk, stalk, and leaves; but a larger proportion of it, and perhaps a better quality, is found in the husk. The attempts to produce paper from this fibre have not thus far been very successful in this country, but in Austria a process has been discovered and patented for making a very superior article of corn-fibre paper, of various grades, and of the finest and strongest texture.

The inventor of this process is Chevalier Auer Van Welsbach, a native of Austria, and a member of the Imperial Government. His experiments have been conducted for a series of years under the patronage of the government, and have resulted successfully in rendering the fibre of maize entirely capable of conversion into paper of all kinds, as well as cloth.

A variety of samples in our possession seem to establish, beyond any doubt, the excellence of this paper, and the fitness of corn-fibre for producing it. It is confidently asserted that the cost of making it from this material is less, compared with the quality, than from any other material known. From the finest tissue to the strongest hardware paper, every intervening grade has been produced by this Austrian process.

It has been officially stated that, on the authority of artists and literary institutions, it is shown that from no other material, so far known, official, drawing, or tracing papers of such durability and tenacity, at equally low prices, have been produced. It is also asserted that the better qualities of post, fancy, and colored papers made of this fibre compete successfully with the finest of the same kind made from rags.

It is also a remarkable fact that, from the same fibre of corn that is found capable of producing this diversity of papers, various grades and textures of cloth have been made, from the thin fabric used for summer clothing to the strongest oil-cloth.

It seems a strange and almost incredible thing, that a plant grown in this country to greater extent and perfection than anywhere else, should be first applied to new and valuable uses under a European invention. Yankee ingenuity, so long proverbial throughout the world, has in this instance been thrown in the shade, and will need to look to its laurels.

One thing is certain: if these fabrics can be produced, by the Austrian process, at the prices and of the qualities claimed for them, which there seems no reason to doubt, it is clearly the interest of this country to have the invention applied on a large scale among the cornfields of the West. Whenever the maize plant shall be made to produce largely, and at a moderate and paying price, other articles of utility and value besides food, it will undoubtedly give a new impulse to the growth and affluence of the country.

SYRUP AND SUGAR.—It has long been known that syrup can be made from the stalks of maize, and recently it has been ascertained that it may be successfully produced from the grain. Various attempts have been made to convert this syrup into sugar, but thus far with doubtful success. The syrup made from the stalk of corn is said to be of fair quality, but will probably never be able to compete with that produced from the Sorghum, now very generally and widely cultivated for the purpose.

There is reason to believe, however, that the syrup produced, by a late invention, from the grain of the corn plant, will be able to compete successfully with most others in the market, in regard to quality and price. This syrup is the product of the starch of corn, and may be made from that element more readily and less expensively than from the grain itself. It is found that a bushel of corn will yield three gallons of the syrup, and the quality is by good judges pronounced excellent.

DISTILLATION.—This cereal has also, like some other of the best gifts of the Deity, been perverted to base and injurious uses. In Ohio and some other parts of the West it is employed in the manufacture of high wines and whiskey. While man is endowed with a twofold nature of good and evil, it is hardly perhaps to be expected that all the beneficent gifts of Providence will be exclusively appropriated to their highest and most valued purposes. But though the amount of corn consumed by the distiller appears large in the abstract, it is yet relatively small, and dwindles to comparative insignificance when viewed in connection with the vast quantities absorbed by other and better uses.

OIL.—The vegetable oil contained in the grain of Indian corn is capable of separation by chemical means, and when thus extracted is more or less useful in various ways. For illuminating purposes it has been tried in some of the light-houses on the Western lakes, and found available. It is doubtful, however, whether the proportion of oil yielded by corn (sixteen gallons to one hundred bushels of grain), taken in connection with the expense of separating it, will render it sufficiently economical for general use.

GREEN MANURE.—For soils deficient in vegetable matter, ploughing in green crops is found by experience to be very useful. It supplies the precise material most wanting in such cases, and in quantities that cannot fail to prove effective. Buckwheat and clover have hitherto been more generally employed for this purpose than any other crop, and the effect is invariably good. But green corn when used for the same object can be made to yield a much larger amount of vegetable matter, and is therefore capable of producing a larger result. Farmers have lately given considerable attention to this subject, and some of the results of recent experience go to show that great and almost incredible fertilizing effects may be in this way accomplished, especially in those cases where the condition of the soil requires a large addition of vegetable matter.

FUEL.—In some parts of the West where corn is abundant and easily raised, and fuel is expensive and difficult to procure, farmers have sometimes found it both convenient and economical in winter to use a part of their surplus corn in feeding their fires. In well-wooded countries, and in the vicinity of coal-regions, this practice will probably never become necessary. But there are districts of country in some of the Western States where the distance from coal mines, the extent of the prairie, and the absence of railroads make it difficult to procure either fire-wood or coal at any reasonable price. It is fortunate for the farmer, in such cases, that Indian corn can be produced at such a rate of cost and in such abundance that, after appropriating all that is needed for the wants of his family and the requirements of his stock, he has still an ample supply left to insure a warm and cheerful hearth through the long winter evenings.

There are those who consider this practice objectionable and wrong, and who seem to be shocked at the idea of burning as fuel a commodity so useful and valuable for food. But a little reflection will show how easy it is for the mind to be so warped by early impressions and preconceived notions as to fail in making simple and obvious distinctions. If this grain was designed by Providence for the use of man, it must clearly have been intended that he should so use it as to derive from it the greatest amount of benefit; and the particular way or the number of ways in which he should use it, is entirely a question of circumstances. That the corn which keeps a man from freezing may be just as useful to him as that which keeps him from starving, is a dictate of common sense too plain to require argument.

It is highly probable that the time is not far distant when the cost of producing corn will be so reduced by improved culture and improved varieties, that the use of it as fuel will be much more general and extensive than it is at present; when it will take the place of fire-wood and coal, not merely occasionally and at a pinch, but in many places constantly and systematically. Indeed, it would hardly be extravagant to anticipate the time when farmers remote from railroads and from wooded districts will make it a part of their regular plan to plant not only a field of corn for the granary, but another for the wood-shed.

MATTRESSES.—The husks of corn are frequently turned to a useful account by farmers and others, in making mattresses, for which they are said to answer exceedingly well, and are highly commended by some who have tried them, on the score of economy and durability, as well as comfort.

THE PRODUCT OF ONE ACRE.

The quantity of food that an acre of land is capable of producing is a question of some interest to society, and one that rises in importance as population advances. There is a period in the growth of every people when the number of inhabitants to a square mile produces a demand for food that raises the question of possible supply.

It is true, the alarm at one time created by the theory of Malthus has been dissipated by later and sounder writers, and men are no longer terrified by the apprehension that increasing population will outrun the means of subsistence until the earth fails to feed its inhabitants. The possibility of this event, if it be a possibility, is too remote to give serious concern to the present generation.

Yet it cannot be denied that, in thickly-settled communities, great interests are at stake on the facilities for procuring food, and on the certainty of its supply; and the importance of preserving and increasing the fertility of the earth becomes in every country more and more apparent from year to year as population accumulates.

Even our own favored land of boundless acres and sparse population is no permanent exception to this universal rule. Here, as in older countries, it is found that the value of land rises with the augmented numbers present to consume its products, and the rapid accumulation of mouths to be fed is prophetic of a coming demand for increased productiveness of soil, and more perfect modes of culture.

It is, perhaps, true enough to-day, that no man in Iowa or Nebraska would feel himself to be any poorer, nor would pay any more for his beef and bread, if the ultimate capacity of each acre were less than it is. There is no present necessity of reaching that ultimate capacity, and consequently no concern felt in regard to it. But these facts are transient. The natural increase of population, augmented as it is by constant accessions from abroad, will in the course of time entirely change this condition, and the now unpeopled prairie will swarm with hungry consumers of bread and meat, that will make it expedient for every farmer to husband the affluence of his soil, and test the capacity of his acres.

From these and like considerations, it will perhaps be interesting to examine some of the capabilities of an acre of corn.

For this purpose, let us assume the product of an acre to be one hundred bushels. This, as before shown, will give a yield of stalks equal to four tons.

It was found, in a previous estimate, that one hundred pounds of the stover are equal in feeding to forty-eight pounds of corn; but in order to accommodate this estimate to the views of those who may possibly rate the value of stalks lower than this, let us take one hundred pounds of them as equal to forty-five pounds of the grain; or, in other words, let us suppose that one hundred pounds of the stover will produce the same amount of butter, beef, mutton, etc., as forty-five pounds of corn.

Comparing this with the estimate made on a former page for the cost of beef, it will be found that, when the stalks and grain are fed separately, it requires about seventeen and a half pounds of the former, or eight pounds, very nearly, of the latter, to produce a pound of beef. If we extend the calculation to other products, the general results will be very nearly as indicated in the following table, which gives the weight of grain, and also the weight of stalk, either of which, separately, will produce one pound of each of the products named:

	For 1 lb. of beef.	For 1 lb. of butter.	For 1 lb. of cheese.	For 1 lb. of milk.	For 1 lb. of mutton.	For 1 lb. of pork.
Corn.............	8 lbs.	10 lbs.	5 lbs.	$\frac{1}{4}$ lb.	4 lbs.	8 lbs.
Stover..........	$17\frac{1}{2}$ "	22 "	11 "	$1\frac{1}{2}$ "	9 "	—

Some of these figures vary slightly from the exact proportion, but they are near enough for practical purposes.

Now, in taking the yield of an acre of corn at one hundred bushels, we shall have five thousand six hundred pounds of grain and eight thousand pounds of stover; but as some varieties of corn and some modes of planting would not give this proportion of stalks, the result stated below is calculated for two different yields of the latter, viz., three tons per acre and four tons.

PRODUCT OF DIFFERENT KINDS OF FOOD FROM ONE ACRE OF CORN.

YIELD OF CORN.		Beef.	Butter.	Cheese.	Milk.	Mutton.	Pork.
Grain.	Stover.						
100 bush.	3 tons.	1,042 lbs.	832 lbs.	1,664 lbs.	16,640 lbs.	2,066 lbs.	1,866 lbs.
100 "	4 "	1,157 "	923 "	1,846 "	18,460 "	2,288 "	1,866 "

The amount of pork given in the last column represents the product of grain only. If the stover omitted in that case were converted into mutton, it would give six hundred and sixty-six pounds for three tons of the stalks, or eight hundred and eighty-eight pounds for four tons. The total product of the acre, therefore, in the last column would be, of pork and mutton together, two thousand five hundred and thirty-two pounds in one case, and two thousand seven hundred and fifty-four pounds in the other.

It is important for the farmer to remember, that when he devotes his acres to either of the above products he gets, in addition to these returns, a liberal supply of valuable manure. The total amount of grain and stalk in the above crop is eleven thousand six hundred pounds in one case, and thirteen thousand six hundred pounds in the other. In the consumption of this quantity of provender, the resulting manure would be worth from forty to sixty dollars, and upward, according to the economy and intelligence exercised in the care and use of it.

It is also to be considered that, in the case of mutton, the value of the fleece is to be added, as part of the product of the acre; and in the case of butter and cheese, the value of the pork made from the refuse of the dairy is, in like manner, a part of the acreable product. It has been estimated that, with good management, the milk of a cow will produce a pound of pork for every pound of butter.

But there is another view of this subject that further illustrates the capacity of an acre of corn for contributing to the support of the human family. When the corn meal is converted into bread and other forms of food for the table, it is found that three pounds of the meal produce over seven pounds of bread, probably seven and a half pounds on an average. Omitting the fraction, this will give more than thirteen thousand pounds of corn-bread per acre, at the rate of yield assumed above. In addition to this, the product of the stover, supposing it to be three tons per acre, would be equivalent to either of the following, viz., to three hundred and forty-two pounds of beef, two hundred and seventy-two pounds of butter, five hundred and forty-four pounds of cheese, five thousand four hundred and forty pounds of milk, or six hundred and sixty-six pounds of mutton.

This would support a family of seven persons for

two years, supposing each man, woman, and child to consume two and a half pounds of bread and one pound of milk per day ; or if, instead of the milk, either of the other products were consumed in the proportion of their yield. But this amount of food to each person is larger than experience has found requisite. Taking the average consumption of food in families, the product of the above acre would exceed the result here given.

CORN CULTURE AT THE WEST.

Agriculture on the Western prairies is quite a different affair, and presents a different aspect, from that with which Eastern farmers are familiar. It is conducted on a scale of such extent, and in a manner so original and peculiar, that it has not only eclipsed all previous ideas on the subject, but seems to have quite bewildered the staid farmers of the older States, some of whom appear to be needlessly disturbed, and imagine all their established theories to be unsettled because the man who plants his corn by the square mile considers it necessary to strike out a theory and practice of his own; as if he imagined his gourd-seed crop of five hundred acres would not be the " big thing " that it is, if raised on the same principles that produce the yellow-flint crop of five acres, or the King Philip of three.

The brave and resolute yeoman who, disdaining his scanty paternal inheritance in New England, has gone forth with a steadfast purpose, and an iron will, to commit his fortunes to the rising West, is naturally impatient of the minute details and commonplace results of Eastern farming, and confidently expects that, in his new quarters, exuberance of soil and multitude of acres will lift him above the drudgery of old methods, make him independent of received maxims, and yet remunerate a minimum of labor with a maximum yield.

If his first crop disappoints him, he is nothing daunted, but plants a wider breadth the following year, still sanguine of success, defiant of chemistry, and superior to the laws of vegetation. He is bound to realize his early dreams of mammoth granaries densely filled; and so long as that end is gained, he cares not to inquire whether it results from quantity of land, or perfection of culture. If when his crop is harvested he can count the product by thousands of bushels, his object is equally accomplished, whether it is the yield of several hundred acres imperfectly cultivated, or of fifty acres thoroughly tilled.

But after all, this passion for doing things on a large scale, at a rapid rate and therefore superficially, is but the outcropping of a national infirmity, and will in due season bring its own remedy. Let us, then, give all due credit to the intrepid pioneer of the prairie, who, though some of his ideas may be more colossal than correct, is yet doing a grand work for humanity, in extending the domain of civilization over an unsubdued wilderness, and transforming the wild pastures of the buffalo into fields of golden grain.

The prevalent notion that the agriculture of the Western States is essentially, and of necessity, a different thing from that of the East, calling for a different set of principles and system of practice, though it may have some slight or apparent foundation in the difference of conditions, is nevertheless pregnant with mischievous error.

Whatever the distinction is, or ought to be, between the modes of culture practised in the two sections of the country, it is supposed to be founded on the fact that in one of these localities land is cheap and labor is dear, while in the other the case is reversed; and also perhaps on the further fact, that where the land is lowest in price, and most abundant, it is, at the same time, the most productive. This at least appears to be the general argument for the Western system.

It does not, however, very clearly appear how the low price and fertile quality of Western land, or even the high price of Western labor, can justify the repudiation of some of the soundest maxims of husbandry. It is not entirely evident that a given amount of corn is more profitably raised from a large area of land than from one of half the extent, even admitting the land to cost less and the labor more than they do at the East. That this may be the case to a limited extent, and in exceptional instances, is very possible. But that it is true in general and in the long run, it would be hard to show. Nor is it easy to perceive the economy of turning cattle into the cornfield during fall and winter to browse on the hard and juiceless remnants of a once nutritious stover from which alternate frost and sun have expelled all the nutriment. Equally difficult also would it be to justify, upon any sound principle, the improvident practice of allowing the cattle and other stock of the farm to enter the cornfield, at the maturity of the grain, to do their own harvesting.

It is urged, in defence of this practice, that the difficulty of procuring help sometimes renders it unavoidable; but if the farmer, instead of planting a half section or more of land, had planted one-half or one-fourth of that extent, adopting at the same time the best modes of culture, he would realize in the end a larger crop, and the labor and expense of harvesting would be greatly reduced. Instead of rambling over miles of territory to gather up a scattered and lean crop, he would have a compact, abundant, and profitable yield within a small compass—a crop that would be easily harvested, and that would pay well for gathering, even at some extra expense for the labor.

The whole of the argument in support of the prevalent system of Western culture seems to be an inversion of the usual mode of reasoning. The superior quality of the soil, so far from being an excuse for careless cultivation, is the best reason in the world why it should be treated in the most thorough manner. It is only the thorough treatment of the land that reveals the fulness of its wealth. The high price of labor, instead of being a reason for diffusing it over a large surface, is an argument for concentration—for bringing it within the smallest compass, where every blow tells, and every stroke is sure of its legitimate effect.

The man who should employ a carpenter at three dollars a day to drive nails, and put him in a position where he could only hit every tenth nail on the head, would be very likely to complain of the expense of labor, but he would hardly assign that as an excuse for his folly; on the contrary, he would find in the price of labor the strongest argument for reforming his practice. Let him place his man in a situation where every blow drives a nail home, and he will not consider the work dearly paid for, whether it costs one dollar a day or three.

It is proper, however, to remark, and it is gratifying to know, that this uncalculating and unprofitable mode of husbandry is by no means universal, in any section of the country, and that there are throughout the West many enlightened farmers to whom these strictures have no application. There is, indeed, through all that country, a marked and increasing tendency toward a better system of culture. The progress of recent years proves that the evil complained of is steadily diminishing and disappearing under the influence of diffused intelligence, of the increasing number of farmers' clubs, and of other multiplied facilities and valuable sources of useful knowledge.

In proportion as men advance in reading and thinking, they gradually acquire the habit of getting a larger amount of products from a less amount of land; and our Western farmers are already beginning to discover that a more careful, calculating, and concentrated culture will produce more corn from an acre, *at a less cost per bushel*, and that a more provident mode of harvesting and feeding will give a larger amount of beef and pork from an acre, *at a less cost per pound*.

13*

THE MANUFACTURING INTEREST IN ITS RELATION TO AGRICULTURE.

Nothing would contribute more, and perhaps nothing so much, to the growth and prosperity of the West and South as the extension and increase of the manufacturing interest. The man who converts raw materials into articles of utility, convenience, or luxury, is a creator of values, and is to that extent a useful and valuable citizen in every community. Like the farmer, he creates products that meet the wants and necessities of men, and his presence is not merely important to society, but indispensable to its progress. The manufacturer undoubtedly ranks next in importance to the farmer, and their avocations are in many points strikingly analogous. It may indeed be said that the farmer, in a broad and important sense, is himself a manufacturer, for, like the latter, he is essentially a creator of values.

There is, therefore, between these two departments of industry a close and intimate connection,—a relationship of mutual dependence and reciprocal benefits. There is probably no country on the globe and no condition of society in which the presence of the manufacturer is more needed at this moment than in some of our Western and Southern States.

The farmer is already there, and is doing his work bravely. He is continually accumulating agricultural products, which, if he had a near-by market, would be synonymous with agricultural wealth. When the manufacturer goes into such a community, he supplies a vacancy that is anxiously awaiting him, and which no one but himself can fill. He finds already there in ample abundance that which he most needs, namely, cheap food and a market for his products, and furnishes in return the very commodities most essential to the wants and necessities of those around him. Thus the proximity of the two classes results in the highest possible advantage to each, and the interchange of commodities becomes a mutual benefit and reciprocal wealth.

The more widely you separate the farmer and manufacturer, the more you impoverish them both. The closer the contact in which you place them, the more you increase and render certain the success and affluence of each. Wherever a manufacturing edifice is reared in the West, the result is a wider home market for beef and pork, and a rise in the price of corn. The advent of factory operatives into a new agricultural region assures coming prosperity to the farmer, and the discordant clatter of machinery that shocks the ears of other men is to him the sweetest of music; for it starts the long dormant corn from the crib, gives new activity and interest to butter and beef, and infallibly prognosticates a new top to the Sunday carriage, a silk gown for the wife, a suit of clothes for the little boy, and a new dress for the baby.

Without assuming to determine the true limit of government policy in fostering the various industrial pursuits, it is certainly much to be desired—nay, infinitely important to the highest good of this nation—that the manufacturing interest should keep pace more nearly with the onward march of agriculture. When these go forward with a uniform and *parallel* progress, mutually aiding and enriching each other, and scattering their useful and valuable products broadcast through the land, the highest condition of material prosperity for the whole country is then fulfilled.

On a comparative view of these great interests, it is perfectly clear that every public measure adopted in favor of the manufacturer promotes indirectly, and probably in the long run to an equal extent, the prosperity not only of the farmer but of every other class in the community; and any line of policy calculated to bring these two producing classes into closer proximity, is a benefit to consumers of every class. It not only tends to increase the supply of their products, but the result is a general and pervading *diffusion* of these needful and useful commodities, with much less of the expensive intervention of railroads and steamers. Thus to the consumer the cost of such products is diminished by all the difference of the expense of transportation, while he also derives a further advantage in the facility of procuring them with promptness and certainty.

MARKET PRICE.

THE market price of Indian corn per bushel is to many farmers, and might well be to all, a matter of comparative indifference. Every judicious cultivator understands that, as a general rule, it is against his interest, and in most cases a blind and mistaken policy, to send his corn to a market town to be converted into money at the current quotations. There are, of course, exceptions to this, as to all general rules. There are times when the market price rises to a level that justifies the husbandman in turning some portion of his crop into ready cash. There are also emergencies that occasionally arise in the experience of farmers when it becomes expedient or necessary to realize prompt returns for their corn crop or a part of it, even though it be at a sacrifice.

Such cases, however, are but necessary evils, and under good management will very rarely occur. The true and obvious policy of the prudent farmer is to feed out his corn on his own premises, thus saving the expense of transportation, and returning to his soil

the elements of fertility extracted by the crop. The
most profitable market for corn, and in nearly all
cases the only profitable one, is to be found in the
cattle-stall, the pig-stye, the cow-yard, and the poul-
try-house; not omitting, of course, the family table,
which, though more limited, is, as far as its requirements extend, the best of all markets.

As every farmer, however, is liable occasionally to
find his interest in resorting to a cash market, not
merely for the sale of his corn, but sometimes and
perhaps more profitably as a purchaser, it is a matter
of some interest to keep himself tolerably posted in
regard to the current quotations, and more especially
is this true in reference to some of the other products
of the farm. In the range of prices for all such provisions as corn is used in producing, he necessarily
feels a lively interest, for in these he discerns the real
profit on his corn crop.

The average price of corn in the New York market for the last three years is about one dollar and
twenty cents per bushel. This price having resulted
from the rebellion, is of course exceptional, and cannot be permanent. For the first two months of the
present year (1866) yellow corn has ranged from
eighty to ninety-five cents. For a long series of years
previous to the war the average was not over sixty-
five cents, and for the last forty years, including the
period of the rebellion, the average price is about
sixty-seven cents per bushel.

The average price of corn for 1865, as compared
with several other products, is as follows:

Corn per bushel.	Beef per lb.	Pork per lb.	Butter per lb.	Cheese per lb.
$1.16	11 cts.	12 cts.	40 cts.	16 cts.

The following table, from the *New York Tribune*, gives the average price of beef cattle per pound each year for the last twelve years:

1854,........per lb. 9 c. full.	1860,...........per lb. 8c. full.
1855,.............10c.	1861,.................7¼c.
1856,.............9½c. nearly.	1862,.................7⅜c.
1857,.............10½c. nearly.	1863,.................9¼c.
1858,.............8¼c. nearly.	1864,..................14⅛c.
1859,.............9c.	1865,.................16c.

The *Tribune* estimates the average weight of cattle marketed at seven hundred pounds per head; and, adding the milch cows to the beeves, as they nearly all eventually go to the shambles, the total number is two hundred and seventy-nine thousand, four hundred and thirty-five head of cattle, representing an aggregate of one hundred and ninety-five million, six hundred and four thousand, five hundred pounds of meat, and thirty-one million, two hundred and ninety-six thousand, seven hundred and twenty dollars in money value. According to its estimate as to sheep (average eighty pounds each, at eight cents per pound), the total of mutton is sixty-six million, nine hundred and thirty-eight thousand, six hundred and forty pounds, costing five million, three hundred and fifty-five thousand, and ninety-one dollars.

For the first two months of the present year the

price of corn compares with other leading products as follows:

	Corn per bushel	Beef per lb.	Pork per lb.	Butter per lb.	Cheese per lb.
JANUARY,	90 cts.	10 cts.	11 cts.	34 cts.	15 cts.
FEBRUARY,	85 cts.	$9\frac{1}{2}$ cts.	11 cts.	35 cts.	18 cts.

CONCLUSION.

Throughout the discussion of this subject, it has been a leading object with the author to illustrate the value of first principles, and to convince the farmer that in order to insure the highest success in cultivating his corn, as well as in using it with advantage, thoroughness of treatment is not merely important and useful, but that it is in fact the one indispensable condition, in which all others are included. This, though true enough in other branches of husbandry, is more emphatically so in the case of corn, on account of its remarkable capacity of development. Its sensitive nature feels and responds to every degree of treatment, rapidly unfolding and expanding under the genial influence of care and effort, springing forward at every touch of thoughtful culture, and, when the hand of skilful labor has apparently exhausted its capability of production, still showing that it has a further capacity of yield—only requiring additional labor and thought, and awaiting the approach of a new and higher method of culture.

It has also been the constant endeavor of the writer to render the discussion of this subject as practical as possible, well aware that, without this quality, it could have but little interest or value for the farmer. Yet it should never be forgotten that in many instances sound practical conclusions are more readily arrived at by the aid of theory than in any other way. Indeed, all reasoning from the facts of experience to general conclusions is of necessity more or less theoretical; and however strong the tendency among cultivators to separate facts from theory, repudiating the latter as of little or no value, still it is only by preserving a proper connection between them that the greatest usefulness of each is found, and the most important results obtained.

It must, however, be admitted that the prejudice prevailing among farmers against theoretical investigation is very easily accounted for and perhaps in some measure justified by the extravagant theories too often propounded by speculative writers—theories with scarcely a fact to rest upon, and certainly not entitled to the confidence of sensible men. It is not, therefore, difficult to understand the jealousy and distrust with which this class of speculations are apt to be viewed by agriculturists.

Yet it does not follow, because some writers indulge in vague and shadowy abstractions, dignifying them with the name of theory, that all theoretical inquiry is necessarily unsound and useless. There is probably no principle nor method of investigation that is not liable to misapplication or abuse; but this consideration, while it furnishes good ground for caution in accepting results, is not a sufficient reason why such method of inquiry should be entirely ignored. Though it justifies careful discrimination between true and false reasoning, it does not warrant the rejection of sound conclusions merely because they are theoretically deduced. It often happens, that the theorist, by pushing his investigations in advance, prepares the way for the practical man, rendering his success easy and certain. When practice, therefore, repudiates all theory alike, without discriminating between the true and the false, it deprives itself of much valuable aid, and rejects a portion of the light that illuminates the path to success.

In nearly all the highest achievements of human ability, thought precedes action, and theory is the precursor of valuable practical results. It was theoretical investigation that, a few years since, announced to the world a new planet in the solar system, in advance of its actual discovery; and the practical astronomer might have long swept his glass over the heavens in a fruitless search for the unknown wanderer, had not the speculative mind of Leverrier given to the instrument its true direction.

It is clearly, then, the interest of the farmer to banish from his mind the narrow prejudice that discerns no truth outside of its own traditions, and repudiates all knowledge derived from books. It is clearly the dictate of practical wisdom to remember that the soundness of every theoretical investigation depends on its relation to facts, and that these rise in value and importance in proportion as they are illuminated by theory; that the most perfect husbandry is that in which fact and theory are harmoniously blended, and that the strong right arm on which the farmer confidently relies works out its best results when it executes the intelligent plans of a thoughtful and reasoning mind.

THE END.

A Report of the Debates and

Proceedings in the Secret Sessions of the Conference Convention for Proposing Amendments to the Constitution of the United States, held at Washington, D. C., in February, A. D. 1861. By L. E. CHITTENDEN, one of the Delegates. Large 8vo. 626 pp. Cloth.

"The only authentic account of its proceedings."—*Indianapolis Journal.*

"It sheds floods of light upon the real causes and impulses of secession and rebellion."—*Christian Times.*

"Will be found an invaluable authority."—*New York Tribune.*

The Conflict and the Victory of

Life. Memoir of MRS. CAROLINE P. KEITH, Missionary of the Protestant Episcopal Church to India. Edited by her brother, WILLIAM C. TENNEY. "This is the victory that overcometh the world, even our faith."—*St. John.* "Thanks be to God, who giveth us the victory through our Lord Jesus Christ." —*St. Paul.* With portrait. 12mo.

"A work of real interest and instruction."—*Buffalo Courier.*

"Books like this are valuable as incentives to good, as monuments of Christian zeal, and as contributions to the practical history of the Church."—*Congregationalist.*

Sermons Preached at the Church

of St. Paul the Apostle, New York, during the year 1864. Second Edition. 18mo. 406 pp. Cloth.

"They are all stirring and vigorous, and possess more than usual merit."—*Methodist Protestant.*

"These sermons are short and practical, and admirably calculated to improve and instruct those who read them."—*Commercial Advertiser.*

The Trial. More Links of the

Daisy Chain. By the Author of "The Heir of Redclyffe." Two volumes in one. Large 12mo. 390 pp. Cloth.

"The plot is well developed; the characters are finely sketched; it is a capital novel."—*Providence Journal.*

"It is the best novel we have seen for many months."—*Montreal Gazette.*

"It is marked by all the fascinating qualities of the works that preceded it. * * * Equal to any of her former novels."—*Commercial.*

Chateau Frissac; or Home Scenes

in France. By OLIVE LOGAN, Authoress of "Photographs of Paris Life," etc., etc. Large 12mo. Cloth.

"The vivacity and ease of her style are rarely attained in works of this kind, and its scenes and incidents are skilfully wrought."—*Chicago Journal.*

"The story is lively and entertaining."—*Springfield Republican.*

The Management of Steel, in-

cluding Forging, Hardening, Tempering, Annealing, Shrinking, and Expansion, also the Case-Hardening of Iron. By GEORGE EDE, employed at the Royal Gun Factories' Department, Woolwich Arsenal. First American from Second London Edition. 12mo. Cloth.

"This work must be valuable to machinists and workers of Iron and Steel."—*Portland Courier.*

"An instructive essay; it imparts a great deal of valuable information connected with the making of metal."—*Hartford Courant.*

The Clever Woman of the Fam-

ily. By the Author of "The Heir of Redclyffe," "Heartsease," "The Young Stepmother," etc., etc. With twelve illustrations. 8vo.

"A charming story; fresh, vigorous, and lifelike as the works of this author always are. We are inclined to think that most readers will agree with us in pronouncing it the best which the author has yet produced."—*Portland Press.*

"A new story by this popular writer is always welcome. The 'Clever Woman of the Family,' is one of her best; bright, sharp, and piquant."—*Hartford Courant.*

"One of the cleverest, most genial novels of the times. It is written with great force and fervor. The characters are sketched with great skill."—*Troy Times.*

Too Strange Not to be True.

A Tale. By Lady GEORGIANA FULLERTON, Authoress of "Ellen Middleton," "Ladybird," etc. Three volumes in one. With Illustrations. 8vo.

"This work, which is by far the best of the fair and gifted Authoress, is the most interesting book of fiction that has appeared for years."—*Cairo News.*

"It is a strange, exciting, and extremely interesting tale, well and beautifully written. It is likely to become one of the most popular novels of the present day."—*Indianapolis Gazette.*

"A story in which truth and fiction are skilfully blended. It has quite an air of truth, and lovers of the marvellous will find in it much that is interesting."—*Boston Recorder.*

What I Saw on the West Coast

of South and North America, and at the Hawaiian Islands. By H. WILLIS BAXLEY, M.D. With numerous Illustrations. 8vo. 632 pp. Cloth.

"With great power of observation, much information, a rapid and graphic style, the author presents a vivid and instructive picture. He is free in his strictures, sweeping in his judgments, but his facts are unquestionable, and his motives and his standard of judgment just."—*Albany Argus.*

"His work will be found to contain a great deal of valuable information, many suggestive reflections, and much graphic and interesting descriptions."—*Portland Express.*

The Conversion of the Roman

Empire. The Boyle Lectures for the year 1864. Delivered at the Chapel Royal, Whitehall, by CHARLES MERIVALE, B.D., Rector of Lawford, Chaplain to the Speaker of the House of Commons, author of "A History of the Romans under the Empire." 8vo. 267 pp.

"No man living is better qualified to discuss the subject of this volume, and he has done it with marked ability. He has done it, moreover, in the interest of Christian truth, and manifests thorough appreciation of the spiritual nature and elements of Christianity."—*Evangelist.*

"The author is admirably qualified from his historical studies to connect theology with facts. The subject is a great one, and it is treated with candor, vigor, and abundant command of materials to bring out its salient points."—*Boston Transcript.*

Christian Ballads. By the Right

Rev. A. CLEVELAND COXE, D.D., Bishop of Western New York. Illustrated by John A. Hows. 1 vol., 8vo. 14 full page engravings, and nearly 60 head and tail pieces.

"These ballads have gained for the author an enviable distinction; this work stands almost without a rival."—*Christian Times.*

"Not alone do they breathe a beautiful religious and Christianlike spirit, there is much real and true poetry in them."—*Home Journal.*

Mount Vernon, and other Poems.

By HARVEY RICE. 12mo.

"Fresh and original in style and in thought. * * * Will be read with much satisfaction."—*Cleveland Leader.*